The Postdoc Landscape

The Postdoc Landscape
The Invisible Scholars

Edited by

Audrey J. Jaeger
NC State University, Raleigh, NC, United States

Alessandra J. Dinin
NC State University, Raleigh, NC, United States

ACADEMIC PRESS

An imprint of Elsevier

Academic Press is an imprint of Elsevier
125 London Wall, London EC2Y 5AS, United Kingdom
525 B Street, Suite 1800, San Diego, CA 92101-4495, United States
50 Hampshire Street, 5th Floor, Cambridge, MA 02139, United States
The Boulevard, Langford Lane, Kidlington, Oxford OX5 1GB, United Kingdom

Notices
Knowledge and best practice in this field are constantly changing. As new research and experience
broaden our understanding, changes in research methods, professional practices, or medical
treatment may become necessary.

Practitioners and researchers must always rely on their own experience and knowledge in
evaluating and using any information, methods, compounds, or experiments described herein.
In using such information or methods they should be mindful of their own safety and the safety of
others, including parties for whom they have a professional responsibility.

To the fullest extent of the law, neither the Publisher nor the authors, contributors, or editors,
assume any liability for any injury and/or damage to persons or property as a matter of products
liability, negligence or otherwise, or from any use or operation of any methods, products,
instructions, or ideas contained in the material herein.

Library of Congress Cataloging-in-Publication Data
A catalog record for this book is available from the Library of Congress

British Library Cataloguing-in-Publication Data
A catalogue record for this book is available from the British Library

ISBN: 978-0-12-813169-5

For information on all Academic Press publications visit our website at
https://www.elsevier.com/books-and-journals

Working together
to grow libraries in
developing countries

www.elsevier.com • www.bookaid.org

Publisher: Andre Gerharc Wolf
Acquisition Editor: Mary Preap
Production Project Manager: Mohanambal Natarajan
Designer: Matthew Limbert

Typeset by TNQ Books and Journals

Contents

Contributors   ix
Acknowledgments   xi
Making the Invisible Visible: A Focus on Postdocs   xiii

**1. History and Evolution of the Postdoctoral Scholar
      in the United States   1**
     Keith Micoli and Steve Wendell

     History   2
     Institutional—Grassroots   6
     Disclosure Statement   13
     References   13

**2. A Review of Postdoc Reforms in the United States and
      the Case of the Fair Labor Standards Act Updates
      of 2016   15**
     Adriana Bankston and Gary S. McDowell

     Introduction   16
     A History of Postdoctoral Reform   16
     The Invisible University: Postdoctoral Education in the United States 17
     Postdoctoral Appointments and Disappointments   22
     Meeting the Nation's Needs for Biomedical and Behavioral Scientists 24
     Trends in the Early Careers of Life Scientists   27
     Enhancing the Postdoctoral Experience for Scientists and Engineers   29
     The Postdoctoral Experience Revisited   33
     Summary   35
     Updates to the Fair Labor Standards Act in 2016   36
     How the Fair Labor Standards Act Updates Included Postdocs   37
     Direct Effects of the Fair Labor Standards Act on Postdocs   37
     **Compliance of Institutions With the Fair Labor Standards
        Act Ruling**   38
     **Institutional Response to the Injunction Against the Fair Labor
        Standards Act Updates**   40
     Conclusion   42
     Acknowledgments   44
     References   44

3. **Institutional Support, Programs, and Policies for Postdoctoral Training**   **49**
    Nisha A. Cavanaugh

    Addressing the Employment Challenges of Postdocs   51
    Defining the Postdoctoral Training Period and Implementing
         Individual Development Plans   56
    Career and Professional Development Programming   58
    Alumni Databases: Tracking Postdoc Career Outcomes   63
    Final Thoughts   64
    Acknowledgments   65
    References   66

4. **Postdoc Scholars in S&E Departments: Plights, Departmental Expectations, and Policies**   **69**
    Xuhong Su and Mary Alexander

    Introduction   69
    Potential Reform Efforts   71
    Departments as Focal Points   73
    Research Hypothesis   74
    Research Design   77
    Expected Career Hierarchy in STEM Departments   77
    Analyses   81
    Discussion   84
    Conclusion   86
    References   87

5. **Proactive Postdoc Mentoring**   **91**
    Sarah C. Hokanson and Bennett B Goldberg

    Introduction   91
    What Is Mentoring?   92
    Characteristics and Outcomes of Successful Mentoring
         Relationships   93
    Encountering Challenges—The Complexities of Real-Life
         Research Mentoring in the Academic Social System   96
    Case Studies   100
    Case 1: Discovering Misalignment   101
    Case 2: Role Models   102
    Case 3: The Perpetual Postdoc   103
    Case Studies: Reflection   105
    Application of Evidence-Based Strategies to Support Success   106
    Faculty Mentors   106
    Postdocs   109
    Institutions   112
    Conclusion   115
    Acknowledgments   115
    References   115

6.   Career Coherence, Agency, and the Postdoctoral Scholar   **121**
Karen J. Haley, Tara D. Hudson and Audrey J. Jaeger

Conceptual Framework: Agency and Career Coherence   122
Mixed Methods Research Design   124
Instruments and Participants   124
Coding and Analysis   126
Researchers and Limitations   126
STEM Postdocs Act to Realize Their Career Goals   127
Survey Findings: STEM Postdocs as Agents of Their Own Career
    Development   127
Learning and Acting: The Career Coherence of STEM Postdocs   127
Three Primary Patterns of Agency and Career Goal Clarity   135
Discussion and Implications   138
Conclusions   141
Acknowledgments   141
References   141

7.   European Cross-National Mixed-Method Study on Early Career Researcher Experience   **143**
Montserrat Castelló, Kirsi Pyhältö and Lynn McAlpine

Introduction   144
The European Context as an Influence on Postdoc Career
    Opportunities   145
A European Cross-National Research Program   146
Research-Related Experiences of Early Career Researchers
    Survey   147
Multimodal Interview Protocol   149
Journey Plots   150
Network Plots   151
Writing as a Primary Form of Research Communication   152
Post-PhD Researchers' Writing Perceptions   153
Development of Writing Perceptions   154
The Study: Exploring Postdocs Writing Perceptions   154
Method   154
Data Collection   155
Participants   157
Analysis   157
Results   158
Quantitative Analysis   158
Qualitative Analysis   159
Discussion   165
Writing   165
Strengths and Limitations of the Research Approach   166
Appendix A   169
References   172

8. Postdoc Trajectories: Making Visible the Invisible   175
Lynn McAlpine

Goal of Chapter   176
Context   177
The Evidence I Am Drawing on   178
Study 1 (2006–2016)   179
Study 2 (2014–2015)   180
Study 3 (2015–2016)   180
Study 4 (2013)   181
What Meaning Do I Draw From the Research?   182
What Do Postdocs Actually Do?   182
How Does Their Work Contribute to Building Their Hoped-for
　Careers?   189
Where Do They Foresee Themselves in 5 Years?   193
Still Hope   193
Considering Alternatives   194
Implications: What Might All This Mean for the Future?   195
Appendix A   198
References   201

9. Global Perspectives on the Postdoctoral Scholar Experience   203
*Karri Holley, Aliya Kuzhabekova, Nick Osbaldiston,
Fabian Cannizzo, Christian Mauri, Shan Simmonds,
Christine Teelken and Inge van der Weijden*

Australia   204
Kazakhstan   208
The Netherlands   213
South Africa   217
Reflections on the International Experiences of
　Postdoctoral Scholars   221
References   223

Index   227

Contributors

Mary Alexander, University of South Carolina, Columbia, SC, United States

Adriana Bankston, Future of Research, Abington, MA, United States

Fabian Cannizzo, Monash University, Melbourne, VIC, Australia

Montserrat Castelló, Universitat Ramon Llull, Barcelona, Spain

Nisha A. Cavanaugh, Sanford Burnham Prebys Medical Discovery Institute (SBP), La Jolla, CA, United States

Bennett B. Goldberg, Northwestern University, Evanston, IL, United States

Karen J. Haley, Portland State University, Portland, OR, United States

Sarah C. Hokanson, Boston University, Boston, MA, United States

Karri Holley, University of Alabama, Tuscaloosa, AL, United States

Tara D. Hudson, Kent State University, Kent, OH, United States

Audrey J. Jaeger, North Carolina State University, Raleigh, NC, United States

Aliya Kuzhabekova, Nazarbayev University, Astana, Kazakhstan

Christian Mauri, Murdoch University, Perth, WA, Australia

Lynn McAlpine, University of Oxford, Oxford, United Kingdom; McGill University, Montreal, QC, Canada

Gary S. McDowell, Future of Research, Abington, MA, United States; Manylabs, San Francisco, CA, United States

Keith Micoli, New York University, New York, NY, United States

Nick Osbaldiston, James Cook University, Cairns, QLD, Australia

Kirsi Pyhältö, University of Helsinki, Helsinki, Finland; University of Oulu, Oulu, Finland

Shan Simmonds, North-West University, South Africa

Xuhong Su, University of South Carolina, Columbia, SC, United States

Christine Teelken, V U University Amsterdam, Amsterdam, the Netherlands

Inge van der Weijden, Leiden University, Leiden, the Netherlands

Steve Wendell, University of Pittsburgh School of Medicine, Pittsburgh, PA, United States

Acknowledgments

Many thanks to Belinda Huang, PhD, for her thoughtful guidance and support throughout this project. Thanks also to Tina Irvine for her careful attention to the editing process. And a special thank you to our authors who are deeply committed to this very important topic and agreed to our ambitious timeline.

Making the Invisible Visible: A Focus on Postdocs

During the period 2002–07, almost half (45%) of all PhD recipients in United States institutions of higher education worked as postdoctoral scholars (Hoffer, Grigorian, & Hedberg, 2008). The National Academies (2014) has detailed concerns about the "postdoctoral training system" for decades: a lack of career development, few mentoring opportunities, low compensation and benefits, and unclear and inconsistent terms of service. Due to these poor conditions, postdocs—a group that often feels invisible—need our attention.

According to the National Postdoctoral Association (NPA) (2007), a postdoc is "an individual who has received a doctoral degree (or equivalent) and is engaged in a temporary and defined period of mentored advanced training to enhance the professional skills and research independence needed to pursue his or her chosen career path." While their roles fit similar parameters, postdoctoral scholars may have different titles at different institutions, leading to difficulty in accurately assessing their numbers. While National Science Foundation (NSF) estimates the number of postdocs is between 30,800 and 63,400, the NPA's survey of members identified 79,000. This discrepancy is just one example of postdoc invisibility.

Another example of postdoc invisibility is in extant literature. Although the NPA and the National Academies have offered several reports, scholarly literature about this growing population is limited. This volume fills that void as it brings together the leading voices about postdoctoral scholars in terms of their presence in government, industry, and the academy; their growing issues and concerns; and a clear direction in terms of what practitioners, policymakers, and educators should do to improve the lives of postdoctoral scholars. This volume will appeal to the individuals at institutions who work with postdocs as it frames their role and experience. At the same time, this volume will serve as a guide for policymakers, including presidents and chief executive officers of organizations, who direct policy change. The volume will also appeal to researchers examining this topic, individuals who supervise and mentor postdocs, and the postdocs themselves.

Part-time faculty were the invisible group in higher education for decades. Now, conversations about part-time faculty are common and institutions are making progress in addressing their needs. The postdoctoral scholar experience is similar to the part-time faculty experience, except very little scholarly research and writing has been offered about postdocs. Without attention to this population, organizations will not be able to meet the growing needs of post-docs. Tremendously qualified talent will go unsupported in an invisible often thankless position. To address this gap, this volume includes chapters centered on three themes: the Postdoctoral Landscape; Postdoc Support; and Postdoc Career Literacy, Agency, and Choice.

Chapter 1 by Keith Micoli and Steve Wendell provides a background history of postdoctoral scholars (postdocs) at United States institutions, focusing on the end of 20th and beginning of the 21st centuries. From grassroots post-doctoral associations to national policy leaders, stakeholders worked to define problems, identify solutions, and change an academic culture that had fostered an unsustainable model of postdoctoral training. One major development in the early 2000s was the creation of the NPA. Founded by a group of local post-doctoral association leaders, the NPA became a bridge between the institutions that trained postdocs and the national organizations (NIH and NSF) that funded them. The NPA provided a voice for the tens of thousands of individual postdocs and worked collaboratively with other stakeholders to develop systemic solutions to the challenges faced by all. The chapter draws to attention the need for further collaborative effort from all stakeholders to enhance data collection on postdocs, propagate and enhance current solutions, develop novel strategies, and refocus on areas that have lagged behind in the evolution of postdoc training.

In the next chapter, Chapter 2, A Review of Postdoc Reforms in the United States and the Case of the Fair Labor Standards Act Updates of 2016, Adriana Bankston and Gary S. McDowell review the history of postdoctoral reforms and changes to the postdoctoral position over time, along with summarizing recent pending recommendations, and assess whether implementation has been successful over time. The history of how the updates to the FLSA came to include postdocs, how they were implemented at academic institutions, and how institutions responded to an injunction preventing the updates from taking effect is discussed as a case study to highlight barriers to implementation of proposed reforms at institutions. The chapter concludes with a discussion of ongoing and future efforts for reform.

Who serves postdoctoral scholars on college and university campuses; who addresses their needs? Chapter 3 by Nisha A. Cavanaugh discusses the structure of postdoctoral offices and the pivotal role they play in assisting and developing postdocs. The postdoctoral population has diversified substantially and this chapter addresses how postdoc offices (1) offer postdoctoral training programs and professional development opportunities that cater to a more diverse population, (2) provide resources to educate postdocs about various PhD careers and how to prepare themselves for their future jobs, and (3) continue to improve the quality of the postdoctoral experience through support and policy.

Xuhong Su and Mary Alexander, in Chapter 4, examine departmental factors that shape the likelihood of postdoc appointments and disappointments on career-related issues. This chapter details the 2010 results of a NSF sponsored survey of a nationally representative sample of chairs and heads in STEM fields at research extensive universities. Specifically it addresses general expectations for doctoral graduates to engage in postdoc training in STEM departments, and what are the best practices departments have been using to promote the status of postdoc appointees? Departments exercise significant influences on individual career prospects and success, particularly for those with well-entrenched interest in departments such as postdoc scholars. It is therefore imperative to examine the roles and practices of STEM departments to unravel the dynamics of postdoc appointment and disappointments.

Opportunities for success early in a scholar's career are primarily influenced by whether or not these individuals have access to engaged, positive, and supportive mentoring relationships. In Chapter 5, Sarah C. Hokanson and Bennett Goldberg discuss how postdoc–faculty relationships have influenced the postdoc's career satisfaction and success. The postdoc position is a balancing act in that it serves as a platform for additional training as well as provides an independent environment that can create challenges for postdocs and their mentors. These challenges affect the postdoc–mentor relationship in many ways: (1) Juggling many of their own responsibilities limits the time faculty can commit to career mentoring and professional development; (2) Faculty have limited knowledge and experience of nonacademic careers, even though many of their postdocs will transition into those pathways; and (3) Increased competition for research funding lowers faculty morale and increases the pressure on their trainees. In this chapter, the authors review the researched-based mentoring literature and identify strategies that institutions and faculty can employ to mitigate some of the overarching challenges that negatively impact faculty mentoring practices and the postdoc–faculty relationship. Through case studies positive aspects of postdoc–faculty mentoring relationships are discussed—establishing expectations, clear communication, fostering independence, and creating inclusive research and teaching environments.

In Chapter 6, the authors, Karen J. Haley, Tara D. Hudson, and Audrey J. Jaeger, explore the connections between elements of Magnusson and Redekopp's 2011 model of coherent career practice and agency to better understand how postdoctoral scholars in STEM fields define and take action to realize their professional career goals. An initial survey of 172 postdocs and 19 individual interviews provided examples of how agency functions in the context of advancing career goals. Findings led to the development of three typical patterns relating to agency and goal clarity. These patterns as discussed in the chapter point toward interventions to better support the career development needs of STEM postdocs.

The next chapter, Chapter 7, offers a European cross-national mixed-method study on early career researcher (ECR) experiences written by Montserrat Castelló, Kirsi Pyhältö, and Lynn McAlpine. The authors write about post-PhD researchers who are a highly accomplished group of ECRs in the second stage of their development as researchers (R2) using the European Commission European Framework for Research Careers. To better understand the challenges they have to confront during their trajectory, the authors developed a European cross-national research program (Spain, United Kingdom, Finland, and Switzerland), which is examining ECRs' identity development. A total of 330 participants (United Kingdom = 98; Spain = 198; Switzerland = 34), mainly from Social Sciences responded to survey items open on agency and self-regulation, interest and commitment on research, scientific communication, community support, and burnout. One year later, a subsample of 52 (United Kingdom = 11; Spain = 22; Finland = 10; Switzerland = 9) participated in a multimodal interview to deep in the significant events and networking in their researcher trajectory. The authors discuss the strengths and limitations of their methodological approach and challenges underlying it: (1) the creation of parallel data collection protocols in four languages; (2) the integration of similar conceptual frames in both the survey and the interview protocols; (3) the use of visual methods in the interview; and (4) the creation of robust qualitative analysis procedures across four national teams.

Lynn McAlpine, in Chapter 8, provides insight into the experiences of approximately 70 postdocs by drawing on research conducted in Canada, United Kingdom, and continental Europe. This international scope is important since the postdoc period often involves considerable mobility across cultural and linguistic boundaries. The research uses a narrative methodology and draws on a range of qualitative data sources thus providing multiple windows into experience. Key findings include the following: managing challenges related to developing research independence; demonstrating resilience in dealing with different forms of academic rejection as they sought to build their scholarly profiles; negotiating scholarly work alongside personal life goals (e.g., establishing a life partner, creating a comfortable work–life balance).

Finally, in Chapter 9, a group of researchers representing Australia, Kazakhstan, the Netherlands, and South Africa reflect on their higher education systems; the role of the postdoctoral fellow within the system; and how internal and external influences shape the postdoctoral experience. Karri Holley, Aliya Kuzhabekova, Nick Osbaldiston, Fabian Cannizzo, Christian Mauri, Shan Simmonds, Christine Teelken, and Inge van der Weijden acknowledge that while widespread concerns exist over the experiences and career trajectories of postdoctoral scholars in higher education, these concerns are rarely examined through the lens of a social and cultural context unique to a national system. Postdoctoral scholars do exist in various forms at academic institutions around the world. Understanding their experiences offers insight not only into the nuanced nature of doctoral and postdoctoral work but also the larger question about how various higher education systems engage in a globalized knowledge economy. This chapter examines the postdoctoral fellow's experience in various national contexts.

Postdocs play an increasingly important role in higher education as institutions seek talented, less expensive labor for laboratories and classrooms. These positions provide critical training and professional development for new researchers yet come with a host of challenges. We invited authors from a variety of backgrounds to speak to the experiences of postdocs internationally. Our authors focus on the most critical issues facing postdoctoral scholars. We hope that through this work we can make postdocs both feel less invisible and ameliorate their experiences.

Audrey J. Jaeger and Alessandra J. Dinin

REFERENCES

Hoffer, T., Grigorian, K., & Hedberg, E. (March 2008). *Postdoc participation of science, engineering, and health doctorate recipients*. InfoBrief, National Science Foundation.

National Academy of Sciences, National Academy of Engineering, and Institute of Medicine of the National Academies. (2014). *The postdoctoral experience revisited*. Washington, DC: The National Academies Press.

National Postdoctoral Association. (2007). *What is a postdoc?* Retrieved from: http://www.national postdoc.org/policy-22/what-is-a-postdoc.

Chapter 1

History and Evolution of the Postdoctoral Scholar in the United States

Keith Micoli[1], Steve Wendell[2]
[1]*New York University, New York, NY, United States;* [2]*University of Pittsburgh School of Medicine, Pittsburgh, PA, United States*

Chapter Outline

History	2	Disclosure Statement	13
Institutional—Grassroots	6	References	13

Topics in this chapter include the following: the challenges postdoctoral scholars faced at the turn of the century and the historical context that contributed to their creation; the early recognition and efforts to face these challenges from diverse perspectives and stakeholders; overarching philosophical and logistical concerns; major milestones in the evolution of postdoc training; the benefits and shortcoming of enacted solutions, and overview of persisting challenges and possible new directions.

Founded by a group of local postdoctoral association leaders, the National Postdoctoral Association (NPA) provided a unified voice for tens of thousands of individual postdocs that had been absent from the national discussion. The NPA became a bridge between postdocs, the institutions that trained postdocs, and the national organizations (National Institutes of Health (NIH) and National Science Foundation (NSF)) that funded them. The NPA worked collaboratively with all stakeholders to develop systemic solutions to the challenges faced by all. The efforts of the NPA and other like-minded organizations helped create a definition of the postdoc and provided a dedicated forum for stakeholders that continue to address ongoing issues and develop additional novel solutions.

The chapter draws attention to the need for further collaborative effort from all stakeholders to enhance data collection on postdocs, propagate and enhance current solutions, develop novel strategies, and refocus on areas that have lagged behind in the evolution of postdoc training.

The Postdoc Landscape. http://dx.doi.org/10.1016/B978-0-12-813169-5.00001-X

HISTORY

Although it is unclear exactly when postdoctoral scholars as an entity first came into existence, we can assume it was sometime after the first Doctor of Philosophy degrees were granted in Germany in the mid-17th century. The title of "doctor" existed long before then and was the product of medieval times, and the German system codified this into a more structured and formal program that was essentially an apprenticeship model. This model was borrowed from the training of other professionals, who would ultimately pursue their careers within a guild. The guild system had Masters, Apprentices, and Journeymen, corresponding roughly to Professor, Student, and Postdoc (Keith Garbutt, 2006).

Despite centuries of technological and scientific advances, the basic framework of this apprentice system remains unchanged today and underpins both the challenges and hope for the future of today's postdocs. In this model, a professor trains students who ultimately become professors in their own right. This model works as long as the ratio of professors to students remains low, and the market for new professors grows fast enough to incorporate the newly minted doctorates. The fundamental flaw in the system is obvious, as the career of a single professor can span several decades and produce many academic offspring. However, this system was relatively stable as long as schools continued to expand, and also limited the size of the new PhD population.

The system began to change in the United States following World War II. The success of the Manhattan Project and the efforts of Vannevar Bush resulted in a dramatic shift in the role of scientific research in the US economy and national security. Vannevar Bush was head of the US Office of Scientific Research and Development and was a key scientific advisor to President Harry Truman. Bush's seminal work, "Science The Endless Frontier," was a report that made the argument that scientific research was integral to national defense interests and to the future growth of the US economy (https://www.nsf.gov/od/lpa/nsf50/vbush1945.htm). This report led to the formation of the NSF and the great expansion of the NIH. The NIH humbly began within the Marine Hospital Service on Staten Island, New York, in 1887, and in 1947 had a budget of $8 million. The postwar expansion of research grants through NIH increased its budget to $100 million in 1957 and reached $1 billion by 1966 (A Short History of the NIH, Victoria A. Harden, PhD, NIH Historian https://history.nih.gov/exhibits/history/docs/page_07.html). This enormous investment in research by the US government, with a model of distributing funds through competitive grants, drove expansion of research at US institutions and created a need for a dramatically increased scientific workforce. Recognizing that a well-trained workforce of sufficient size was in the national interest, Congress passed the National Research Act in 1974, which was signed into law on July 12 by President Richard Nixon. This law created the National Research Service Award (NRSA) program and funded both predoctoral and postdoctoral scholars. The law requires that those supported on an NRSA, either as an individual or through an institution, be citizens of the United States or legal permanent residents. The commitment to training a biomedical and behavioral workforce adequate to address the nation's needs was reflected in the fact that the training budget for NIH was 15% of the total NIH budget following passage of the National Research Act. Congress also required that the overall training program be evaluated regularly by an independent group, with the goal of recommended changes to policy and areas of research deemed to be important to the national interest. This committee releases a report of its findings, "Personnel Needs and Training for Biomedical and Behavioral Research," published by the National Academies Press and archived reports can be accessed for every year a report was published (www.nap.edu).

From the earliest reports, the need for better data collection was noted by committee members, an issue that has been noted in every report since, covering a span of over 40 years. Without an accurate assessment of the number of trainees in the workforce, it has been very difficult to propose policies that would adequately address the true needs of the current and future workforce. The most complete and accurate data available does include the NRSA program, making it the de facto national training model despite its limitations (most notably that it does not include any trainees who are in the United States on temporary visas) (https://www.nap.edu/catalog/12983/research-training-in-the-biomedical-behavioral-and-clinical-research-sciences).

The NRSA program in its early years is estimated to have supported approximately 50% of all graduate students and 33% of postdoctoral fellows in biomedical research and helped to maintain a relatively stable production of PhDs at around 3000 per year until the mid-1980s (NRC report 2011, Fig. 3.2). One unintended consequence of this expansion of the scientific workforce was an almost immediate increase in the number of postdoctoral fellows. According to data from the Survey of Doctorate Recipients presented in the NRC report, there were approximately 2000 academic postdoctorates in 1973 and over 4000 in 1979. The number of tenure-track faculty positions remained essentially unchanged over that time, at roughly 22,000. In fact, the absolute number of tenured or tenure-track faculty positions has been remarkably steady and was still below 30,000 as of 2006 (NRC report 2011, Figs. 3–7).

What has changed is the number of nontenure track or "other" academic appointments. There were approximately 2500 nontenure track positions held in 1975, expanding to approximately 25,000 in 2006. This is a 10-fold increase during a time in which the tenured faculty population remained essentially unchanged. Whatever the reason for this dramatic shift in the composition of the more senior ranks of academic research, the effect was the creation of a bottleneck that made tenure-track positions more and more difficult to obtain. This occurred at the same time that total enrollment in biomedical graduate programs was increasing, moving from 30,000 in 1983 to 50,000 in 2008. With no increase in tenure-track positions at one end of the pipeline, and a dramatic increase in PhD production at the other, the postdoctoral position became the main pressure-relief valve in an expanding biomedical research workforce.

The postdoctoral population in the United States expanded dramatically but quietly, as the total extent of this expansion was difficult to quantify. The NSF has been charged with the task of monitoring the research workforce and accomplishes this task primarily through the Survey of Earned Doctorates, the Survey of Graduate Students and Postdocs, and the Survey of Doctoral Recipients. Each of these was designed for a specific purpose and provides important pieces of the overall workforce puzzle. However, all of them were designed during an era in which the expected and most common career outcome for a recent PhD was a tenure-track faculty position. Each survey also has key weaknesses, and thus the total size of the postdoctoral population remains a very rough estimate, currently somewhere between 45,000 and 85,000. This lack of precision is a very important factor to account for when any predictions about future workforce needs are made and hampers any attempts to measure the effectiveness of policy changes aimed at impacting the workforce pipeline.

One potential cause for the troubling lack of data on postdoctoral researchers may be that by its very nature, the postdoctoral period is temporary and was originally seen more as an optional period of dedicated research before embarking on an independent career. Prior to the mid-1990s, there were no administrative units devoted specifically to postdoctoral fellows at US research institutions. This population fell into the cracks between graduate students, faculty, and staff, and oversight fell to the individual faculty members who hired, trained, and mentored the postdocs.

Concerns about postdoctoral training and initial steps to identify solutions were growing by the late 1990s in pockets across national organizations, within institutions, and at grassroots levels with postdocs themselves. One of the primary challenges was simply how to identify postdocs since institutions typically lacked a recognized definition for postdocs. Definitions varied between institutions and even within institutions where a wide array of job classifications including some "catch-all" classifications were used (at least 24 job classifications used at a single institution including "volunteer"). This presented a major hindrance to obtaining even rudimentary information about the number of postdocs, compensation and benefits, and training environment.

This growing concern was becoming clear at both the national level by the mid-1990s, and three key reports would be published by the close of the 20th century that highlighted the need for change, and set key recommendations for improving the scientific training system in the United States. The 1994 report, *Meeting the Nation's Needs for Biomedical and Behavioral Scientists* (National Research Council Committee on National Needs for Biomedical and Behavioral Research Personnel, 1994) marked the 20th anniversary of the formation of the NRSA. This report acknowledged that the employment landscape for scientists had changed and made several recommendations to modify the NRSA accordingly:

1. Stipend levels for both predoctoral and postdoctoral trainees should be increased to $12,000 and $25,000, respectively, and increase by 3% annually.
2. The number of awards should remain at current levels in basic biomedical research fields.

3. Assist women in establishing themselves in productive careers as research scientists.
4. Improve data collection on NRSA recipients and their career outcomes.

This report is also notable in acknowledging both the shift in apparent career outcomes and the demographics of the scientific workforce: "Employment conditions for biomedical and behavioral scientists were relatively robust throughout the 1980s. Dramatic changes have occurred, however, with regard to sector of employment with a greater fraction of Ph.D.s employed in industry and other nonacademic jobs than in earlier years. The nation's need for research scientists has also been affected by demographic changes: the number of individuals from racial and ethnic minority groups is increasing but not as fast as might be expected given federal efforts to encourage the participation of minorities in this area. The work force of the future will consist of an increasing proportion of women and minorities; it is important that these changes are reflected in the biomedical and behavioral science work force." (*Meeting the Nation's Needs for Biomedical and Behavioral Scientists* (National Research Council (US) Committee on National Needs for Biomedical and Behavioral Research Personnel, 1994)).

This report was limited in scope to the NRSA program itself, though the committee acknowledged the disproportionately large effect the NRSA has on the training system despite supporting a diminishing percentage of trainees over time.

A report published in 1998, Trends in the Early Careers of Life Scientists (National Research Council (US) Committee on Dimensions, Causes, and Implications of Recent Trends in the Careers of Life Scientists, 1998), addressed the the growing "crisis of expectation" that grips young life scientists who face difficulty in achieving their career objectives.

The findings of the Trends in the Early Careers of Life Scientists report included an increase in age at receipt of PhD (32-years old, on average), a more than doubling in the percentage of new PhDs who pursued a postdoctoral position (from 21% in 1973 to 53% in 1995), increased time spent in postdoctoral training (5 years), and increased age at time of first permanent position (increasing by 8 years for tenure-track faculty positions over 1973). What was most alarming to the committee, and was indeed the impetus for its inception, was the sharp decrease in successful NIH grant applications from individuals younger than age 37. This trend troubled those who worried that the most productive and creative years of a scientific career were now being spent in an apprenticeship position that had now become a "holding pattern."

Among its recommendations, the Trends report included the following:

1. Constrained growth in PhD production.
2. Dissemination of accurate information on career prospects.
3. Shifting support of PhD students from research grants to training grants and fellowships.

4. Enhance independence of postdoctoral researchers through career transition awards.

5. The PhD remains a research-intensive degree, with the primary purpose of training independent scientists.

This final recommendation was made despite recognition that employment in academia was stagnating, and that a growing percentage of new PhDs was finding employment in "alternate" careers. The committee acknowledged that career aspirations were shifting among some students and that graduate programs should adjust to expand awareness of growing career opportunities. However, the committee felt that current preparation for nonacademic careers was insufficient and that competition for these careers was intense. This tepid acknowledgement at least began the discussion of what remains a divisive topic today and laid the groundwork for further progress.

The national discussion on the state of scientific training reached a critical stage in 2000 with the report *Enhancing the Postdoctoral Experience for Scientists and Engineers, by the Committee on Science, Engineering and Public Policy (COSEPUP)* (*Enhancing the Postdoctoral Experience for Scientists and Engineers:* A Guide for Postdoctoral Scholars, Advisers, Institutions, Funding Organizations, and Disciplinary Societies (2000). This report remains the most influential guide to national reform efforts and laid out more than 50 recommended changes for individuals, institutions, funding organizations, and disciplinary societies. The COSEPUP report represented the most holistic approach to date and brought together many stakeholder groups that previously had worked in isolation on these issues.

Science magazine hailed the report in its story, Shadow People No More: New Report Puts Postdocs in the Light (http://www.sciencemag.org/careers/2000/09/shadow-people-no-more-new-report-puts-postdocs-light). This article also announced the creation of The Postdoc Network (PDN), an evolution of Science's *Next Wave*, with funding from the Alfred P. Sloan Foundation. PDN became an incubating space where like-minded individuals could exchange ideas, build relationships, and connect with others. The COSEPUP report and the creation of PDN tied together the efforts of national organizations and the local, grassroots efforts underway at many institutions.

INSTITUTIONAL—GRASSROOTS

Even prior to the *Enhancing the Postdoctoral Experience for Scientists and Engineers, by the COSEPUP* report in 2000, some institutions were taking actions to address postdoc training. There was significant variation among institutions regarding who raised the concern, what initial efforts were undertaken, and the subsequent path and pace of change. In some institutions, proactive administrations formed committees or tasked individuals with fact finding and providing recommendations. At another level, individual or groups of faculty within institutions served as primary champions for changes to postdoc training.

Postdocs themselves also began to raise concerns and sought to contribute to improvements.

Postdoc offices (PDO) or officers began to become established at institutions that varied widely in terms of staffing, budget, mission, and administrative functions. This included offices representing little more than a faculty member with a percent effort to organize a professional development workshop series to offices with multiple staff charged with significant oversite including human resources functions such as postdoc appointment approval. Groups of postdocs began to form postdoc associations (PDA) or similar councils. The first known PDA appears to have begun in 1992 at Johns Hopkins School of Medicine (COSEPUP, 2000). Although priorities varied, common early efforts included establishing professional development resources, career option resources, and institutional policy changes related to compensation and benefits, and more consistent job classification usage.

While the COSEPUP report outlined actions for institutions including establishing explicit policies, PDO, and PDA, there was a lack of ongoing support to help navigate this implementation. The recommendations also included actions that could be taken by funding agencies and professional societies, which highlighted the need for ongoing dialogue across these organizations. This dialogue should also include multiple perspectives from the institutional elements including the postdocs themselves. While important contributions came from groups like the Association of American Medical Colleges, Group on Graduate Research, Education, and Training (AAMC-GREAT), which includes individuals that provided leadership for postdocs at institutions, it was becoming clear that an organization solely dedicated to postdoc training was needed to bring together all stakeholders, including postdoc themselves.

The American Association for the Advancement of Science (AAAS) Next Wave PDN annual meeting provided the first forum at a national level to bring stakeholders together to focus on postdoc training. It was at the second annual PDN meeting in April 2002 (http://www.sciencemag.org/careers/2002/01/announcing-2002-postdoc-network-national-meeting) that seven postdocs decided that the postdocs needed a national voice and inspired the creation of the NPA. The founding and early development of the NPA itself reflects the potential of collaborative efforts by different stakeholders represented by the various early mentors and advisors to the NPA leadership (http://www.sciencemag.org/careers/2003/03/national-postdoc-association-makes-its-debut). Among the other attendees at the PDN meeting were additional postdocs that were themselves leaders or founders of PDAs at their institution and would join the fledgling NPA on the first elected executive board or make contributions on committees. Members of institutional PDOs, funding agencies, professional societies, and organizations such as AAMC-GREAT, and AAAS were early mentors and committee members in the NPA. Michael Teitelbaum, as program director at the Alfred P. Sloan Foundation was an important advisor and was pivotal in securing funding to make the NPA possible. Staff members at the

AAAS served as principal investigators on a Sloan Foundation grant, and the AAAS further agreed to house the nascent NPA while also providing continued mentorship (Postdocs: A Voice for the Voiceless, Nature 489, 461–463 (2012) https://www.nature.com/naturejobs/science/articles/10.1038/nj7416-461a). This ensured that many of the stakeholder perspectives and the collective expertise were engaged during the foundation of the NPA. However, the first elected NPA executive board were all postdocs, ensuring that the mission, strategy, and persona of the NPA were reflective of the national postdoctoral community. The importance of the NPA leadership embracing the perspectives of all stakeholders, appreciating the complexity of the scientific enterprise, and reaching out broadly for partnership cannot be overstated. These have been critical to the NPA's ability to play an important role in advancing postdoctoral training. This is particularly important in context of the ideas presented in this book that expand the perspectives around postdoc training.

The NPA bridges postdoc training stakeholders on multiple levels, including (1) facilitating relationships within the institution such as administrators, PDO, and postdoc representatives in the PDA, (2) facilitating ongoing communication between national entities including funding agencies, professional organizations, and governmental policy makers, and (3) connecting the national level dialogue and institutional stakeholders affected by policies and charged with implementing solutions. An important example of the NPA's impact is the support for building, maintaining, and evolving the postdoctoral training structures at institutions. Toolkits for starting and maintaining PDO and PDAs have provided valued advice that has contributed to their widespread and rapid creation. In addition, the ongoing sharing of best practices and innovative strategies facilitates timely and broad dissemination to the larger postdoc community. The NPA's annual meeting contributes to an active national dialogue among stakeholders while also disseminating collective advice, best practices, and innovative solutions. Data collection efforts, such as the NPA's Institutional Policy Database, represent a resource that provides value to stakeholders at institutions for benchmarking, while also contributing to the national dialogue. The NPA has also drawn on the different stakeholders to contribute to common recommendations such as the Core Competencies and white papers such a White paper to the NIH NPA White Paper to the National Institutes of Health (c.2003) that articulated the need for creation of a K award for postdocs that is strikingly similar the eventual NIH Pathway to Independence Award, K99/R00 (https://grants.nih.gov/grants/guide/pa-files/PA-15-083.html). The K99/R00 award provides support for 2 years of postdoctoral research and up to 3 years during an independent research track faculty position. This ensures financial and scientific independence during postdoctoral research and portability for the transition into a research faculty position. In essence, one might say that the NPA combines many parts that lead to value greater than the sum of its parts.

These critical functions of the NPA to provide a voice to the national dialogue, facilitate data collection, monitor institutional policy and practices, support the continued growth and health of institutional postdoctoral support structures, disseminate novel approaches, and facilitate institutional implementation are highly relevant to the concerns and unmet challenges outlined in the chapters of this book. The NPA represents the predominant way in which individuals and groups responded to the challenges facing postdoctoral scholars, but this collaborative approach is not the only one being used. From its inception, the NPA was forced to address the notion of a postdoctoral union as a means to bring change. The NPA officially remains neutral on the issue of whether postdocs should seek collective bargaining rights under the umbrella of a union.

The first known example of postdocs forming a union was at the University of Connecticut in 2003, roughly coinciding with the formation of the NPA. This was a critical time in the evolution of postdoctoral training, and it was not at all clear that institutions would be more inclined to heed the warnings of the numerous reports previously described, and frustrations among postdocs were high. The fact that unionization was becoming a national discussion was probably an underrecognized driver of change, even though few institutions saw significant attempts to unionize their postdocs. What the University of Connecticut postdoc union achieved was far more important nationally than locally, and its establishment forced institutional administrators to weigh the risks of not changing against the risks of unionization. This led to the "soft pressure" for institutions to make the admittedly overly simplified choice of working with groups such as the NPA to create positive change in a collaborative way or see their postdocs seek change through collective bargaining. The evidence of this decision-making process is seen in the enormous growth in creating administrative units devoted to postdocs, from fewer than 30 in 2003 to over 150 in 2015 (NPA institutional policy, http://www.nationalpostdoc.org/?policy_report_databa). Postdoc unions also sought to alleviate the inequities in the postdoc training system and focused on salaries, vacation time, and grievance/termination policies, although career development resources became a bigger focus more recently. The NPA also highlighted the need to address issues of salary and unregulated human resource policies but placed more emphasis on the career development needs of postdocs. By far the largest, and best known, postdoc union is in the University of California (UC) system. The International Union, United Automobile, Aerospace and Agricultural Implement Workers of America (UAW) represents over 6500 postdoctoral researchers in the UC system. This system-wide union resulted in equalizing postdoctoral researcher pay and benefits across schools that, until then, had shown a wide range of policies and practices regarding postdocs. Some UC schools, such as UC San Francisco and UC Berkeley, were among the national leaders in postdoc policies and had substantial career development programs already in place. By unionizing, the postdocs within the UC system were brought under unified policies. The benefits included salaries tied to the NIH NRSA stipend scale, defined holidays and vacation time, defined procedures for termination and layoffs, and a grievance procedure to protect the rights of postdocs. The UAW postdoc union serves as a safeguard against the worst abuses in the training system, but it is not likely designed to address the deeper issues of the system itself nor provide support for institutional investment in creative new professional development programs beyond those that already existed.

A successful postdoctoral experience ends with an individual no longer being a postdoc. That transition is at the heart of the problems facing our training system, and it is far more complex than how much money a postdoc is paid, how many vacation days he or she receives, or how long he or she is allowed to be called a postdoc. Unions play an essential role in many industries to protect the interests of their members. However, a union cannot easily come up with a solution to the very personal nature of the relationship between a faculty member and a postdoc. The union can protect the worker, but they cannot influence the job market available to a postdoc or the degree to which a faculty member supports their career decision.

Unions, or perhaps the threat of unionization, have no doubt helped spur progress on the issues facing postdocs. The NPA, by working at multiple levels collaboratively, has also spurred progress. Each approach brings benefits and limitations. Unions have brought standardized policies to institutions that had not only resisted such things for decades but also may have furthered the "us versus them" attitude that many postdocs hold. The NPA helped organize postdoctoral offices, which allowed implementation of many of the COSEPUP recommendations, but in the process of doing so has been perceived as representing the interests of institutions over the individual postdocs they endeavor to help. It is important to note that another group, Future of Research (FoR), was more recently founded as a grassroots postdoc organization in the Boston area in 2014. Since their initial symposium, FoR has worked to promote data transparency, better data collection, and empowerment of postdocs to work for systemic change. The efforts of FoR are reminiscent of the NPA's founding and represent another avenue to creatively solve the entrenched problems in the current scientific training enterprise.

Change has been slow and incremental; yet, the accretion of change over the past 15 years is both significant and heartening. It might be argued that those changes were long overdue and/or were the low-hanging fruit easiest to reach, but the foundation laid by this progress is necessary to future change.

In addition to multiple perspectives on policy-based approaches by national funding agencies, ideas for better alignment of postdoc numbers with academic careers have been discussed (https://www.nature.com/naturejobs/science/articles/10.1038/nj6929-354a, http://www.pnas.org/content/111/16/5773.full.pdf). These include restricting the pipeline at the graduate level either through reduced matriculation, transitioning training support to be more dependent on training grants or fellowships, or actively preparing graduate students to choose careers outside the academic path to reduce the number that seek postdoc positions.

An example of the latter was presented at the first NPA meeting in 2003 where keynote speaker Dr. Keith Yamamoto's "Hub with Spokes" model (http://www. sciencemag.org/careers/2003/03/national-postdoc-association-makes-its-debut) viewed graduate programs as the Hub to prepare graduate students for the many different career options or spokes within academia and beyond.

There are multiple challenges to efforts to restrict the pipeline including the dependency on graduate students and postdocs as the research engine. As previously mentioned, postdocs are also cheaper and more experienced than graduate students so even if the pipeline were restricted in the United States, the incentives exist to fill positions with international talent. Another approach for aligning postdoc positions with academic careers is to create, or increase, staff scientist positions as an academic career choice that could replace some of the positions previously filled by a turnover of multiple postdoc over time. These positions would no longer be viewed as temporary training but as valued academic career choices. The details of these positions are critical and harken back to a comment made during the first NPA meeting reflecting the concern that we address changes to postdoc training in a way that we do not push postdocs into a different job category and have a need for a "Research Associate Association" (or staff scientist) in 10 years. It is not unheard of some new job categories at institutions to be viewed as "superpostdocs." While the idea of incorporating more of the postdoc population into staff scientist positions has benefits, it is worth acknowledging the overall shift occurring within academic careers in the 21st century that mirrors the private sector at some level. The traditional lifelong employment and notion of climbing the ladder have shifted to a more transient nature as employers are responding to volatility and the increased pace of change by increased reliance on jobs with less long-term commitment. This is represented in academics by the growth in number of nontenure track faculty positions and adjunct faculty.

On a macro level, the strategies and policies required to achieve a systemic change to the postdoc numbers and condition are incredibly complex. What is sometimes underappreciated in the discussion are many of the subtleties at the micro level of decision-making entities including, individuals, departments, and institutions. Entities making decisions such as whether to create a new graduate program are often not considering the big picture that we already train too many graduate students in some disciplines. While the best of intentions often fuel these decisions, we need to acknowledge that at some level, political considerations can have a strong influence such as the prestige of starting a new graduate program. Similarly, what are the responses if an existing graduate program is forced to shut down considering at least some individuals have a vested interest in the program? Another example that could be argued is occurring currently around the initiation of internship programs for postdocs. There is a noble purpose and reasonable premise behind these efforts but depending on the details of the internship it is easy to imagine how the expansion of using internships for postdocs as a career exploration resource could do harm to the system.

Remembering postdoc training originated as an internship-like or apprentice experience with related rationale, there is potential for an industry internship to become the new expectation for postdocs in an "arms race" of credentials and experience. Eventually could this become a requirement for certain sectors of industry? Other efforts to help postdocs could have the same impact such as formal business and leadership training especially if they become a separate industry with certifications and special relationship with industry. The difficulty comes from the macro versus micro viewpoint. The micro view of a postdoc and postdoc offices is understandably centered on what will benefit current postdocs and help them stand out (at least for now) in competing for a job rather than considering the impact on the macro level. This is further complicated by the vested interests at multiple levels for creation of new graduate programs, resistance to reduction of graduate programs in certain disciplines, and development new programs that elevate the credential expectations.

As previously discussed, many of the changes to postdoc training in the early 21st century were important but more logistic in nature. These have laid the critical foundation for the next level of systemic changes such as proposals to restrict the pipeline. While these systemic changes are further considered and implemented, there has been an increasing effort to better prepare both graduate students and postdocs for the various career options they will pursue. This is a major current focus and represents preparation for offboarding at both the graduate and postdoc levels while also creating the optimal benefit under the current postdoc system.

This has led to effort by funding agencies, professional societies, and institutions at multiple layers to develop resources to explore career options, enhance career-specific skills sets, and increase career planning. Additionally, the availability of supporting resources available online is staggering. We now have many resources that were unimagined 20 years ago while more are continuously being explored.

One of the challenges that has become apparent is the underutilization of resources such as career exploration, career development, and the individual development plans. We view these resources as necessary but insufficient. We need to address strategies to enhance the affective, behavioral, and cognitive capacities of graduate students and postdocs that are transformative in their engagement with available resources. We also know metrics that are predictive of career outcomes that can more quickly measure the impact of interventions. The Career Adapt-Ability Scale (CAAS) has considerable potential (Savickas & Porfeli, 2012). This validated 24 item instrument is the subject of considerable research for association with objective and subjective career outcomes (Rudolph, Lavigne, & Zacher, 2017). Recent intervention strategies using certified professional coaching (International Coach Federation) have proven successful in significant enhancement of graduate students and postdoc CAAS scores as well as additional parameters such as motivation for their scholarly projects and strength of career preference (Wendell, unpublished data).

Exploring new intervention strategies such as professional coaching and metric such as the CAAS to assess intervention strategies represents new avenues that unlock graduate and postdoc engagement with the innovative professional and career development resources that continue to expand.

In summary, the current challenges facing postdoctoral scholars and the institutions that support their training have evolved from an academic system that was not designed or intended to include large numbers of postdocs. The ad hoc nature of the academic training system and unregulated growth culminated in the late 20th century with the recognition that engaging all stakeholders in the training enterprise would be necessary for successful change to occur. Despite the fact that many challenges remain, the landscape today is brighter than it was 20 years ago, thanks in large part to the many groups and individuals who contributed their effort and expertise toward driving positive change. The organizations and collaborations that contributed to these changes have ongoing challenges outlined in the chapters of this book to ensure postdoc training fully honors the investment of public funds. The future of postdoctoral scholars in the United States is unclear, but it is certain that understanding the history of scientific training will be crucial in shaping the next step along the way.

DISCLOSURE STATEMENT

Dr. Steven Wendell's research on supplemental professional coaching (International Coach Federation accredited) is funded by a Harnisch Grant from the Institute of Coaching at McLean Hospital, Harvard Medical School Affiliate. In addition, he is the owner of Fortitude Coaching LLC with financial interests derived from teaching the ADAPT Career Coaching elective course in the Professional Coach Certification Program at Duquesne University.

REFERENCES

Committee on Science Engineering and Public Policy (U.S.) (2000). Enhancing the postdoctoral experience for scientists and engineers: A guide for postdoctoral scholars, advisers, institutions, funding organizations, and disciplinary societies. Washington, DC: National Academy Press.

Garbutt, K. (2006). CBE life sci educ. *Spring*, *5*(1), 39–40.

Rudolph, C. W., Lavigne, K. N., & Zacher, H. (2017). Career adaptability: a meta-analysis of relationships with measures of adaptivity, adapting responses, and adaption results. *J. Vocat. Behav.*, *98*, 17–34. http://dx.doi.org/10.1016/j.jvb.2016.09.002.

Savickas, M. L.Porfeli, E. J. (2012). Career Adapt-Abilities Scale: Construction, reliability, and measurement equivalence across 13 countries. *Journal of Vocational Behavior*, *80*(5), 661–673.

Chapter 2

A Review of Postdoc Reforms in the United States and the Case of the Fair Labor Standards Act Updates of 2016

Adriana Bankston[1], Gary S. McDowell[1,2]
[1]*Future of Research, Abington, MA, United States;* [2]*Manylabs, San Francisco, CA, United States*

Chapter Outline

Introduction   16
A History of Postdoctoral Reform   16
  The Invisible University:
  Postdoctoral Education in the
  United States   17
    Overview   17
    Recommendations   19
    Analysis   20
  Postdoctoral Appointments and
  Disappointments   22
    Overview   22
    Recommendations   23
    Analysis   23
  Meeting the Nation's Needs
  for Biomedical and Behavioral
  Scientists   24
    Overview   25
    Recommendations   25
    Analysis   26
  Trends in the Early Careers of
  Life Scientists   27

    Overview   27
    Recommendations   28
    Analysis   28
  Enhancing the Postdoctoral
  Experience for Scientists and
  Engineers   29
    Overview   30
    Recommendations   30
    Analysis   32
  The Postdoctoral Experience
  Revisited   33
    Overview   34
    Recommendations   34
    Analysis   34
Summary   35
**Updates to the Fair Labor
Standards Act in 2016**   36
**How the Fair Labor Standards
Act Updates Included Postdocs**   37
**Direct Effects of the Fair Labor
Standards Act on Postdocs**   37

The Postdoc Landscape. http://dx.doi.org/10.1016/B978-0-12-813169-5.00002-1

15

Compliance of Institutions With the Fair Labor Standards Act Ruling	38	Conclusion	42
Institutional Response to the Injunction Against the Fair Labor Standards Act Updates	40	Acknowledgments	44
		References	44

INTRODUCTION

The nature of the postdoctoral position has long been debated. The specific role of a postdoc within an institution, what the postdoc is, what the postdoc should be, and even how many postdocs there are, are all discussions that are complicated by the multitude of titles given to postdocs nationwide today. A number of reports over the last decades starting in 1969 have proposed reforms to the postdoctoral position (Anon, 1981; Committee to Review the State of Postdoctoral Experience in Scientists and Engineers et al., 2014; Curtis, 1969; National Academy of Sciences, National Academy of Engineering, Institute of Medicine, Committee on Science, Engineering, and Public Policy, 2000).

In this chapter, we summarize the history of major reports targeting the postdoctoral and highlight successes and failures in reform to the postdoc position. As an example, we analyze one case study examining changes to postdoctoral salaries at the national level to investigate whether postdoc reforms have been or can be successfully implemented.

A HISTORY OF POSTDOCTORAL REFORM

While it is difficult to know exactly when and where postdoctoral researchers came into existence (a roughly equivalent fellowship system was set up at Johns Hopkins in 1876 (Johns Hopkins University, 1914)), one formal mechanism established in 1919 was the National Research Fellowship Program by the National Research Council, funded by the Rockefeller Foundation, which ran for three decades (Anon, 1919). Following the dramatic expansion of the research enterprise during and after the Second World War (which includes the "Golden Years of NIH Expansion" (National Institutes of Health, 2016a)), both the number of postdoctoral researchers and the expectation that the professoriate would undertake postdoctoral training prior to faculty appointment increased (Figure 1 and Supplementary Figure 1 in Heggeness et al. (2016)). Mention of postdocs increased from 1944 (Israeli, 1944) and in reports from the early 1960s, including a small survey commissioned by the Association of American Universities in 1962 (Berelson, 1960, 1962). By 1969, the postdoctoral position had become a sufficient source for concern about the value of its role in academia as well as a focus for recommendations to lead to the first comprehensive national study and subsequent publication of the first report on the postdoc position by the National Academy of Science, "The Invisible University" (Curtis, 1969).

The Invisible University: Postdoctoral Education in the United States (Curtis, 1969)

The present report is the result of a concern within the National Research Council and elsewhere about the scope of postdoctoral education in the United States. Although postdoctoral appointees were present on many campuses, their numbers and functions were not known nationally and, in many instances, were not even known to the host universities. Postdoctoral education, as the title of this report suggests, had grown to institutional status without study or planning. In the absence of information, the costs and benefits of this development to the universities, to the postdoctoral appointees, and to the nation could not be adequately assessed. The financial uncertainties associated with reductions in the federal research budget during the last several years added to the urgency of the need for information.

The Invisible University: Postdoctoral Education in the United States (Curtis, 1969).

Overview

This report claims to be the first comprehensive study of "the postdoctoral" and is the beginning of our review of recommendations. The report is prefaced by a description of the postdoc position, which follows: Postdocs *"will seek to work full time in research for a year or two with a senior investigator,"* while *"the mentor of such young [people] finds them almost indispensable"* in that the lab is managed by the postdoc. With this arrangement, it was proposed that the mentor can carry out teaching and committee work, and the postdoc can bring fresh new ideas. The preface of the report discusses the incredulity with which the authors were met in preparing it: Why study postdocs when *"postdoctoral education was perhaps the best part of higher education?"* But it was the absence of study and the need for data and information about the postdoctoral position that the report cited as the very reasons the report was necessary, to provide a mechanism for discovering and addressing problems with the postdoctoral position.

The 1969 description of the postdoc position sounds similar to the present day, with some slight differences. As the report states, *"The postdoctoral scholar with a PhD is most often a young natural scientist who has recently completed [their] doctoral dissertation,"* yet there were other postdocs at the time in the social sciences and humanities, and there were also postdocs with MDs. Interestingly, more senior researchers on sabbatical also fell under the category of postdoc: *"A good percentage of the postdoctoral population consists of faculty members who have taken leave from their institutions to study in a colleague's laboratory or in a library that offers resources they need."* This situation differs from the current description of the postdoc, which is usually expected to take place immediately after postgraduate study for the PhD, or after residency training following the MD.

The 1969 report discusses the unclear titles given to postdocs, including the overlap with "research associates" and the confusion over the term "postdoctoral fellow," which was typically used for postdocs on fellowships, as seen in the report's statement: "*A postdoctoral scholar's status is not always clear from [their] title.*" This confusion persists today, when 37 titles have been identified in use for postdocs at US institutions by the National Postdoctoral Association (McDowell, 2016; Ferguson et al., 2014).

The 1969 report also discusses the close association of postdocs with "distinguished institutions" and claims that "*although postdoctorals can be found at almost 200 universities, over half of them are at only 17 institutions.*" In 2015, data from the National Science Foundation's (NSF) Survey of Graduate Students and Postdoctorates in Science and Engineering (GSS) suggest that there are 344 US institutions with postdocs in science, engineering, and health; out of the 63,861 total postdocs identified by the survey effort, 31,944 (or 50%) can be found at 27 institutions (Table 46, 2015 (National Science Foundation, n.d.-a, n.d.-b, n.d.-c)). Therefore, this suggests that over the span of nearly 50 years, 50% of postdocs have remained at approximately 8% of US institutions.

Anecdotally, large lab sizes were observed in 1969 as they are today: "*Characteristic also is an association between post-doctorals and distinguished mentors. It is not difficult to find internationally known investigators serving as mentors to as many as a dozen postdoctorals.*" Another phenomenon observed in 1969 that rings true today is the international character of postdocs: "*Approximately 55 percent of post-PhD's and 40 percent of postprofessional doctorate recipients in universities are not U.S. citizens… A very large proportion of the total population is foreign… Most people involved with postdoctorals are aware of the fairly large numbers of foreign citizens within the group… Implicit in all of these attitudes and concerns are questions concerning the numbers of foreign postdoctorals.*" 52% of the US biomedical workforce in 2014 was foreign-born (Heggeness et al., 2016, 2017), and 55% of postdocs were reported to be on temporary visas in 2015 (National Science Foundation, n.d.-a, n.d.-b, n.d.-c). The 2012 Biomedical Workforce Working Group Report suggests that the proportion of noncitizen postdoctoral researchers has grown since 1980 (Biomedical Research Workforce Working Group, 2012), and a comparison of data between 1967 and 1979 implied a decrease in that time of the percentage of foreign postdoctorals from 45% to 38%, with varying fields showing increases or decreases in the percentage of their postdocs who were foreign (Figure 6.6; Anon, 1981).

The report estimates the postdoctoral population in 1967 to be 16,000: "*That this large a number of holders of the doctorate should be welcome at several hundred different host institutions,*" the report states, "*implies that something is very right about postdoctoral study. The eagerness with which former postdoctorals are sought by university departments for faculty positions suggests that the experience and/or the selectivity of the postdoctoral appointment makes this group particularly attractive.*" This report, therefore, highlights the fact that the postdoc position was considered relatively valuable. Despite this fact, many of the issues faced by postdocs in 1969 are largely the same as today.

In 1969, *"both the participants and the subsequent employers seem to consider postdoctoral education a success"* and the report goes on to say, *"For almost a decade, university presidents have been concerned about the ever increasing number of postdoctoral appointments on campus."* Despite this concern, the number of postdoctoral researchers has continued to grow from 1979 to the present day, aside from a bubble of expansion from 2008–10, which corrected from 2010 to 2013 (previously identified as the end of post-doctoral expansion (Garrison, Justement, & Gerbi, 2016)); however, growth appeared to resume after this period (National Science Foundation, n.d.-a, n.d.-b, n.d.-c).

The nature of the postdoc position and how it became so poorly defined perhaps is summed up by the statement that *"few universities have initiated postdoctoral activity by design. When asked why his university encourages post-doctoral education, one graduate dean replied: 'I am not sure we could be said to have a rationale; we permit rather than promote postdoctoral study.'... Unlike undergraduate and graduate education, postdoctoral education is, with few exceptions, not consciously or intentionally undertaken by the university."* The issues debated about postdoctoral researchers are strikingly familiar nearly 50 years later. Indeed, this first 1969 report specifically about postdocs makes note of comments in a prior report on graduate education from 1960, which dedicated 10 pages to postdocs, including the statement that *"there is so much post-doctoral training that many people are becoming perplexed or even alarmed at where it is all going to end"* (Berelson, 1960).

Recommendations

- *"All those connected with postdoctoral education are urged to conceive of the postdoctoral appointee as one who is in the process of development and not primarily as the means to accomplish other ends."*
  - The report recommends that funding agencies recognize that research support can be an explicit benefit to an institution's educational goals.
  - The report recommends that institutions recognize the educational benefit that postdocs provide to the institution.
  - The report recommends that principal investigators provide every opportunity and encourage postdocs to be independent investigators.
- *"Because of the individual nature of personal development, we believe that the participation of the post-doctoral in administration and teaching and the duration of the appointment should be determined in each individual case."*
  - The report points out that many postdocs engage in and are interested in, teaching, administration, or other academic pursuits. It therefore recommends tailoring individual postdoc development toward these goals.
- *"Current restrictions should be removed to allow postdoctoral fellows to choose mentors at industrial research laboratories."*

- *"The allotment of existing space and the planning for new facilities should include explicit recognition of the anticipated postdoctoral population at both the immediate and senior levels."*
- *"Postdoctoral fellowships should carry with them sufficient support for research expenses, so that the fellow need not depend on his mentor's sources of support to carry out his proposed research."*
- *"The number of postdoctoral opportunities available at any time should be related to the number of Ph.D.'s and professional doctorate holders who can profit from the experience."*
  - The report recommends a careful distribution of numbers/proportions of "Fellows," i.e., postdocs on individual fellowships, "Training Associates" (i.e., postdocs on departmental training grants), and "Research Associates" (i.e., postdocs on research grants), in particular, to help those requiring training for their individual aspirations.
- *"Support for senior and intermediate postdoctoral opportunities should be increased in all fields."*
  - The report refers to a substantial number of postdocs who have either returned to academia from employment elsewhere for further training ("intermediate" postdocs) or faculty taking paid sabbatical positions ("senior postdocs").
- *"[T]he foreign postdoctoral is a most welcome visitor... This exchange of persons can be stimulated by cooperating in programs that are designed to encourage the foreign postdoctorals to return to their homelands."*
- *"Travel of American postdoctorals abroad should be encouraged and the number of opportunities increased."*

Analysis

The most striking reflection on reading "The Invisible University" from 1969 is how similar today the discussion about postdocs was almost 50 years ago, although some individual recommendations have been put into place. The largest differences between the 1969 report and the current postdoc situation are related simply to numbers and demographics, but it is, again, apparent that many of today's trends were evident even 50 years ago. The report itself repeatedly refers to the *laissez-faire* attitude toward postdocs in 1969, which persists generally in academia to this day.

One of the greatest differences between postdoctoral populations in 1969 and today is the length of postdoctoral training, which in 1969 was 1–2 years depending on the field, and today is reported to be 5 years on average (Kahn & Ginther, 2017). However, current postdoctoral training may actually be longer, as at least 30% of postdocs are reporting undertaking at least two postdoc positions (Powell, 2012), and as individual postdocs are not tracked, this implies that 5 years may be simply the average for an individual postdoc position, but not representative of how long individuals are actually undertaking postdoctoral research.

Another big difference is the career outcome following postdoc training—in 1969, 20% of biology PhDs would go straight to faculty positions, likely to be a rare occurrence today. In addition, the demographics of the 1969 postdoc population were overwhelmingly male and, presumably, also white. The proportion of international researchers was roughly half, whereas today it is estimated to be around two-thirds. There is no concerted effort to track their movement, and studies of the foreign researcher are limited (Franzoni, Scellato, & Stephan, 2015; Stephan & Levin, 2001). Most indications are that foreign postdocs tend to stay in the United States rather than leave (Franzoni et al., 2015; Heggeness et al., 2016), and the number of US PhDs carrying out postdoctoral research abroad is very small (Polka, 2014).

The educational benefit which a postdoc provides to an institution, particularly through mentoring more junior scientists, is still arguably largely unnoticed. The ability of postdocs to act as independent investigators can be extremely variable. Recently, there have been movements toward greater individual postdoc development, with implementation of tools such as myIDP (Independent Development Plan, http://myidp.sciencecareers.org (Fuhrmann et al., 2011)) and programs such as the Institutional Research and Academic Career Development Awards from the National Institute of General Medical Sciences (NIGMS) (IRACDA, https://www.nigms.nih.gov/Training/CareerDev/Pages/TWDInstRes.aspx) to train postdocs in teaching. Access to individual career development training varies with institution and principal investigator and is an issue of concern to current junior scientists, who, in many cases, have developed their own career development opportunities.

Compared to academia, postdoctoral programs in industry are mostly independent of academic postdoctoral efforts. Therefore, it is unclear whether the restrictions on postdocs (to which the 1969 report alludes) have actually been lifted, rather than industry professionals developing their own parallel programs. Likewise, institutions have expanded space and research capacity to accommodate postdocs, particularly during the "NIH doubling," where the budget of the NIH was doubled over the period of 1998–2003 (Alberts, 2010).

The number of postdocs nationally has continued to expand, and as postdocs themselves are poorly tracked, this suggests that any changes in numbers and proportions of postdocs in different categories must have occurred indirectly as a result of other factors such as funding and research capacity decisions and in an uncontrolled manner. Essentially, all postdocs today are what the report describes as "immediate" postdocs or those who have transitioned directly from their PhD into the postdoc position. The intermediate postdoc is not a widely discussed/observed phenomenon, and senior postdocs or sabbaticals are often a property of individual institutional arrangements, not in an arrangement such as a postdoctoral appointment.

Postdoctoral Appointments and Disappointments (Anon, 1981)

For many of the most talented scientists and engineers the postdoctoral appointment has served as an important period of transition between formal education and a career in research. The appointment has provided the recent doctorate recipient with a unique opportunity to devote his or her full energies to research without the encumbrance of formal course work or teaching and administrative responsibilities. Those holding such appointments have made valuable contributions to the quality, creativity, and productivity of ongoing scientific inquiry. While the overall magnitude of these contributions has varied markedly depending on the field of research, postdoctorals have played a significant role in the research effort in virtually every field of science and engineering--even in those fields in which their numbers have been quite small. Whether or not the postdoctoral will play an important role in the future, however, will depend on how universities and the scientific community as a whole adapt to a rapidly changing environment.

Postdoctoral Appointments and Disappointments (Anon, 1981).

Overview

This report aimed to provide the next comprehensive study of the postdoc meant to reassess the situation laid out in "The Invisible University" (Curtis, 1969). Contrary to "The Invisible University," "Postdoctoral Appointments and Disappointments" takes on a more urgent and concerned tone, pointing out from the very beginning that concerns have begun to arise about the post-doc position. In particular, they highlight concerns over *"the lack of prestige and research independence in postdoctoral appointments for the most talented young people,"* as well as a gap in the importance of the postdoc and the limited opportunities available for those with postdoctoral training on completion of their postdoctoral research combined with a lack of recognition of postdocs in university (which was the main inspiration for the title of the 1969 report, "The Invisible University," highlighting the lack of recognition of postdocs). In contrast to the 1969 report, the lack of participation by women and underrepresented minorities in postdoctoral research is a major concern in the 1981 report.

The report was compiled including four surveys: one of chairs of science and engineering departments with postdocs in 1977; one of US citizens who received science or engineering doctorates from July 1, 1971 to June 30, 1972; one of a similar cohort but from 1977 to 1978; and one of foreign postdocs.

To define the postdoc, the report used the definition of, *"a temporary appointment the primary purpose of which is to provide for continued education or experience in research usually, though not necessarily, under the supervision of a senior mentor. Included are appointments in government and industrial laboratories which resemble in their character and objectives post-doctoral appointments in universities."* Of note in this definition is that mentoring by a more senior researcher appears optional. The estimates of the number of postdocs in science and engineering range from 8411 to 13,856, or roughly twofold estimate of postdoctoral numbers that persists to this day (Biomedical Research Workforce Working Group, 2012).

One of the concerns was that the rate of faculty hiring had declined in the years prior to the report, citing National Research Council data that from 1969 to 1977, faculty increased by 3% per annum, whereas for the 8 years prior, the increase had been at 10% per annum. The report also cites an NSF survey carried out in 1973 asking postdocs what their most important reason for taking the post-doc position was. Twenty-five percent of those who had received their PhD in the last year had cited that no other employment option was available, a statistic that increased to 37% for those 2 years out of their PhD, and 46% for those 3 years out. The report states that *"It seems that many postdoctorals, like planes waiting to land, were stacked in a holding pattern,"* 32 years before Henry Bourne's use of a similar metaphor of the postdoctoral "holding tank" (Bourne, 2013).

Recommendations

- The report recommends establishing federally funded portable fellowships: 250, plus 50 exclusively for underrepresented minorities, for scientists and engineers, having competitive 2-year stipends and research expense funds.
- The report recommends that each university *"with sizable numbers of nonfaculty research personnel"* establish committees explicitly focused on postdocs and other nonfaculty research staff to review their position and recommend institutional policies.
- The report urges the NSF to expand their survey efforts to look at career decisions of young scientists and engineers.

Analysis

With regards to federally funded portable fellowships, the National Science Foundation does currently have postdoctoral fellowships that generally follow the postdoc. For example, the Directorate for Biological Sciences administers the Postdoctoral Research Fellowships in Biology, which address three areas: Broadening Participation of Groups Under-Represented in Biology (which goes toward the recommendation to ensure there is a particular proportion ear-marked for underrepresented groups); Research Using Biological Collections; and National Plant Genome Initiative Postdoctoral Research Fellowships. There are about 40 fellowships in this category and other fellowships distrib-uted among other directorates. The Broadening Participation of Groups Under-Represented in Biology awards were begun in the 1990 financial year, nearly a decade after the recommendations of the 1981 report were made. These awards are for 2–3 years and include research expenses: for the 2017 application cycle, a monthly stipend of $4500 (or annual salary of $54,000) is awarded, with allow-able costs for fringe benefits, and $15,000 "research and training allowance" to be "spent at the Fellow's discretion," which covers conference travel as well as various research expenses (National Science Foundation, 2017). Therefore, this recommendation appears to have been addressed.

However, the other recommendations appear not to have been addressed comprehensively, and at least not in response to the 1981 report. The issues with postdoctoral administration and policy have been slow to develop, and measures to address the postdoc position at institutions appear to have waited until the formation of the National Postdoctoral Association, which pushed to establish postdoc offices and policies addressing postdocs at institutions. Those efforts began more than 20 years after the publication of the 1981 report and continue to the present day (National Postdoctoral Association, n.d.).

There is still no survey examining the career decisions of junior researchers. Since the 1981 report, no new surveys have been implemented (the closest, the Survey of Doctorate Recipients (National Science Foundation, n.d.-a, n.d.-b, n.d.-c) was already in place at the time of the 1981 report, and the authors were requesting a survey reaching beyond this). The forthcoming Early Career Doctorates survey (National Science Foundation, n.d.-a, n.d.-b, n.d.-c), again, only addresses those currently at academic institutions and does not address the pressing need to find where those who were previously at these institutions have gone.

At the end of the Abstract, the report states, *"Beyond these specific recommendations, the committee believes that the entire postdoctoral institution is in a state of transition and must be reexamined by federal and university policymakers."* It would not be until 1994 that another report mentioned postdocs in the context of biomedical science (National Research Council (US) Committee on National Needs for Biomedical and Behavioral Research Personnel, 1994), and it would take nearly 20 years for the next study dedicated to postdocs (National Academy of Sciences; National Academy of Engineering; Institute of Medicine; Committee on Science, Engineering, and Public Policy, 2000).

Meeting the Nation's Needs for Biomedical and Behavioral Scientists (National Research Council (US) Committee on National Needs for Biomedical and Behavioral Research Personnel, 1994)

In 1994 we mark the twentieth anniversary of the National Research Act of 1974 (P.L. 93-348), which established the National Research Service Awards (NRSA) program... We cannot emphasize too strongly the significant impact the NRSA program has had on the federal system of predoctoral and postdoctoral training at U.S. universities... the National Research Act of 1974 established a coherent system of support for recruiting individuals into health research and launching them into productive careers. Coupled with a variety of mechanisms to support training and education at all stages of the scientific career—from high school through midcareer—NIH provides the largest research training effort in the federal government, the centerpiece of which is the National Research Service Award... The continuing need for highly trained specialists to conduct research to meet the health needs of the country is as great today as it was 20 years ago.

However, because of changes in patterns of research funding and the structure of the marketplace, the nature of this need has changed somewhat in recent years. Today there is a greater demand than in the past for talented health scientists to provide leadership in industrial research settings, in federal government laboratories, and in hospitals and clinics. The NRSA program continues to play a critical role in the preparation of many of those scientists.

Meeting the Nation's Needs for Biomedical and Behavioral Scientists (National Research Council (US) Committee on National Needs for Biomedical and Behavioral Research Personnel, 1994).

Overview

The purpose of this report was to provide recommendations based on the experience provided by 20 years of the NRSA, which provides stipends to graduate students and postdoctoral researchers.

The report highlights recent changes in the biomedical labor market, stating that *"employment conditions for biomedical and behavioral scientists were relatively robust throughout the 1980s. Dramatic changes have occurred, however, with regard to sector of employment with a greater fraction of Ph.D.s employed in industry and other nonacademic jobs than in earlier years."* While this may be true to some extent today, preparation and training of PhDs for employment in sectors other than academia is still largely lacking across the United States.

The report also highlights changes in demographics, which are not as swift as anticipated: *"The nation's need for research scientists has also been affected by demographic changes: the number of individuals from racial and ethnic minority groups is increasing but not as fast as might be expected given federal efforts to encourage the participation of minorities in this area. The work force of the future will consist of an increasing proportion of women and minorities; it is important that these changes are reflected in the biomedical and behavioral science workforce."* As with career preparation, this issue of lack of diversity in academia still persists today.

Recommendations

- The report recommends the NIH raise postdoctoral NRSA stipends: "Postdoctoral NRSA awardees [are] earning approximately $18,600 in their first year of training and $19,700 in their second. It becomes very difficult at this important period of training to entice a clinician or Ph.D., already burdened with debt, into research training. Thus, the committee recommends that the NRSA stipends at the first-year postdoctoral level be increased to $25,000 in inflation-adjusted dollars by fiscal 1996."
  - The report recommends that the yearly NRSA postdoctoral stipend should be raised to $25,000 by fiscal year 1996.
  - The report further recommends that the stipend should increase each year with inflation by 3%.

- The report has a number of recommendations regarding the specific numbers of certain NRSA awards:
  - Maintain the annual number of postdoctoral awards in the basic biomedical sciences at the 1993 level or approximately 3,835 awards.
  - Increase the annual number of NRSA awards for research training in the behavioral sciences to 500 by 1996.
  - In the Medical Scientist Training Program (MSTP), *"increase the number of postdoctoral NRSA fellowship awards from 68 in fiscal 1993 to 160 by fiscal 1996 to permit the preparation of patient-based investigators. To permit the expansion of the pool of MSTP trainees and postdoctoral fellows, we believe modest reductions should be made in the number of postdoctoral awards made through institutional training grants in the clinical sciences. NIH reports that 2,051 awardees were supported in fiscal 1993 through this mechanism. We believe a gradual decrease in the number of awards to 1,805 should occur by fiscal [year] 1996."*
- The report expresses concern that women may be disadvantaged in the NRSA mechanism or subsequent receipt of NIH grants. Therefore, the report recommends more research training opportunities for women and strengthening postdoctoral support to assist women in achieving independence, particularly in allowing for greater flexibility around family concerns.
- The report recommends better data collection of postdoc training outcomes and emphasis, particularly in data collection around minority groups participating in NRSA funding mechanisms.

Analysis

The subject of postdoctoral salaries has come up in many reports, and the 1994 report is the first clear recommendation for salary raises. The NRSA minimum stipends were, in fact, raised in 1994, but only to $19,608, which remained the same in 1996. In fact, the first-year postdoctoral NRSA stipend did not exceed $25,000 until 1999 (National Institutes of Health, n.d.). NRSA stipends also have neither increased annually nor in regular 3% increments since this recommendation. Instead, they have been largely constant for several years at a time, and then increased at rates of 1% or 2% annually in the 2004–12 period. In other years, they have leapt by as much as 25% in 1 year, such as between 1998 and 1999. The NRSA stipend raises have, therefore, not been made in a sustainable manner over time but have likely occurred in response to factors other than the recommendation for a sustainable increase in stipends.

A recurring recommendation appearing consistently throughout several reports, including the 1994 report, is that of tracking outcomes following postdoc training and improved data collection for this purpose. This initially appeared in the "Postdoctoral Appointments and Disappointments" report of 1981 in the form of a specific recommendation to survey career interests and directions of junior researchers (Anon, 1981), which is still yet to be implemented. This recommendation will continue to appear across reports to the present day; therefore, while it is a priority from the point of view of those recommending it, does not appear to be a priority from the perspective of those responsible for its implementation.

Trends in the Early Careers of Life Scientists (National Research Council (US) Committee on Dimensions, Causes, and Implications of Recent Trends in the Careers of Life Scientists, 1998)

The committee recognized that it was dealing with interdependencies among educators, trainees, investigators, funders, and entrepreneurs that truly constituted a sociotechnical system of great complexity. The importance of established stakes in the status quo quickly became apparent, and the committee recognized that there was no single locus of power to make changes in the system that has produced undesirable outcomes for some young scientists. If change is to occur, it will be through the uncoordinated action of many persons at many institutions who try to consider what is best for their students and their profession and then take appropriate action. Those insights tempered any ambition that the committee might initially have had to 'reform' the system overnight by taking bold measures. The risk of doing more damage than good is great, given the complexity of the educational system, the size of the enterprise, and its importance for the nation's long-term interest. Accordingly, the committee's principal recommendations are measured rather than dramatic.

Trends in the Early Careers of Life Scientists (National Research Council (US) Committee on Dimensions, Causes, and Implications of Recent Trends in the Careers of Life Scientists, 1998).

Overview

This report focuses particularly on graduate students and postdocs, and, in contrast to the 1994 report, is a wider assessment of the postdoctoral position, albeit focused on the life sciences (as many reports on postdocs tend to be). The report again cites employment prospects and labor trends as sources for concern about the entry of junior scientists into life sciences.

The report finds that since the 1960s and 1970s, junior scientists by 1998 were slightly older when beginning graduate school and were also taking 2 years longer to graduate, which occurred at the average age of 32. In addition, PhD holders were twice as likely to take a postdoc position, and by this time, their number was estimated at 20,000 in the life sciences. Finally, the time spent in training had increased, and the postdoc "holding pattern" of researchers aged 35–40 in this position, which persists to this day, was established.

Recommendations

- The report calls for greater transparency and greater distribution of correct information about the career outcomes of junior scientists in the life sciences.
  - The report calls for up-to-date career prospects to be made available.
  - The report also suggests that programs should publicly post the outcomes of junior scientists from their departments and provides prospective graduate students with the outcomes of students from the last 10 years of the program.
- The report calls for more opportunities for postdocs to develop independence.
  - The report recommends that all funding agencies should establish a pool of grants that facilitate "career transitions" or transitioning to independence.
  - The report recommends that postdoctoral research positions should not be available to a junior scientist for longer than 5 years.
- The report discusses the path to nonacademic careers in the life sciences.
  - However, the report asserts that the PhD degree should remain a research-intensive degree.
  - The report states that nonacademic careers should be "discussed but not oversold" to graduate students.
  - The report points to Master's Programs as a way of directing junior scientists toward nonacademic careers in areas for which they suggest Master's degree training may be more appropriate.
- The report recommends that foreign researchers still be admitted to early career positions, and that while the number of visas should not be constrained, the number of researchers enrolling in these programs should.

Analysis

Transparency in career outcomes from training programs is still a topic of discussion today (Polka, Krukenberg, & McDowell, 2015), and some institutions have only recently begun to release these data (Silva et al., 2016). The report calls for transparent reporting of the career outcomes of previous junior scientists to then facilitate the ability of current junior scientists to make rational career decisions. There has been little success in widespread dissemination of career outcomes, particularly for postdocs, which are still poorly tracked.

The degree of independence postdoctoral researchers possess is still a topic of current debate; indeed, there seems to have been a downward trend in the independence described by postdocs in the 1969 report, "The Invisible University." Independence was largely taken for granted (although some cases of exploitation, particularly of foreign postdocs, are described in "The Invisible University"), but is an increasing cause for concern conveyed in both the 1998 and subsequent reports (Committee to Review the State of Postdoctoral Experience in Scientists and Engineers et al. 2014; National Academy of Sciences et al., 2000; National Research Council (US) Committee on Dimensions, Causes, and Implications of Recent Trends in the Careers of Life Scientists, 1998).

The recommendation of the 1998 report about discussing nonacademic careers is particularly weak and this discussion continues today, although now there is a more forceful call for presenting an accurate picture of the career landscape. The recommendation that nonacademic careers should not be "oversold" (particularly to graduate students) may mean that there was no effective reduction in the number of graduate students pursuing postdoctoral positions. This is likely due to a lack of effective communication of the realities of the job market combined with the need for effective and early career decision points. There has been some development in the meantime to push Master's programs as a means to facilitate an earlier career decision point and to possibly encourage junior scientists to pursue careers for which a Master's degree may be more appropriate. However, anecdotally in the United States, in contrast to Europe, Master's degrees tend not to be viewed favorably. In particular, when Master's degrees are discussed, a primary objection raised is that this does not provide the labor that the biomedical enterprise is dependent on.

The report also discusses that foreign nationals are critical to research and that international movement is desired in science (Franzoni et al., 2015; Stephan & Levin, 2001). The report is careful to point out that visas should not be limited, but that enrollment of foreign researchers in early career positions should be. Foreign researchers make up the majority of the biomedical workforce (Heggeness et al., 2016, 2017), and the number of foreign researchers willing to take on postdoctoral positions in the United States exceeds the number of positions available at any time. This necessarily leads to a tension in many discussions in terms of how to reduce the number of postdocs, which will not decrease without intervention due to the improved prospects which the postdoc position can provide to foreign nationals.

Enhancing the Postdoctoral Experience for Scientists and Engineers (National Academy of Sciences; National Academy of Engineering; Institute of Medicine; Committee on Science, Engineering, and Public Policy, 2000)

Postdocs have become essential in many research settings. It is largely they who carry out the sometimes exhilarating, sometimes tedious day-to-day work. Their efforts account for a great deal of the extraordinary productivity of the United States' academic science and engineering enterprise. And yet the institutional status of postdocs, especially in academia, is often poorly defined. Consequently, although most postdocs value highly their experiences and the opportunity to engage in rewarding research without competing responsibilities, many of them are dissatisfied with their situations.

Enhancing the Postdoctoral Experience for Scientists and Engineers (National Academy of Sciences, National Academy of Engineering, Institute of Medicine, & Committee on Science, Engineering, and Public Policy, 2000).

Overview

This is the first truly comprehensive report specifically looking at postdoctoral researchers across the scientific enterprise since "The Invisible University" in 1969. The 2000 report points out that the status of postdocs in institutions is ill-defined, unclear, and uncertain, which was the same observation made 31 years prior to this report. This report points out that funding agencies supporting post-docs are unable to provide the number, salaries, benefits, or many other data points on postdocs. The report also shows that the rapid expansion of postdocs has led to a wide range of salaries, some of which are undesirably low, especially when considering postdocs with families to support and those in high cost-of-living areas.

The report, like many others, states that change must be collaborative, and that while postdocs themselves play a role, change ultimately requires action from those in power: namely, *"the advisers, the research institutions, and the funding organizations."*

Recommendations

- The report defines a set of guiding principles:
  - That the postdoc is an apprenticeship experience for gaining scientific, technical, and professional skills to advance their career;
  - That postdocs should receive recognition for their work and compensation, including fringe benefits;
  - That postdoctoral appointments should have a clear purpose, mutually agreed upon by the postdoc, the adviser, the institution, and the funding agency.
- Action points are directed at advisers, institutions, funding organizations, and scientific societies:
  - Award institutional recognition, status, and appropriate compensation;
  - Develop policies specifically for postdocs;
  - Develop greater communication with postdocs;
  - Annually evaluate postdocs;
  - Ensure access of postdocs to health insurance and all institutional services;
  - Set term limits for the postdoc position;
  - Include postdocs in relevant decision-making processes about postdoc policies;
  - Provide career advice to postdocs;
  - Improve collection and dissemination of career data on postdocs;
  - Take actions to improve career transitions for postdocs.

Recommendations are then organized by the various stakeholders:

- Postdocs should
  - Take responsibility for undertaking a postdoc and informing themselves about career options, institutional resources, and policies;

- Contribute their best efforts to the program in which they work;
- Share with their advisers the responsibility for frequent communication;
- Bear the primary responsibility for the success of their experience.
- Advisers should
  - Ensure that the postdoc is provided with an educational experience;
  - Put down in writing their expectations for the postdoc;
  - Mentor the postdocs;
  - Annually discuss/evaluate postdoc performance;
  - Provide career counseling/job placement assistance for postdocs;
  - Assess whether multiple postdoc mentors are required.
- Institutions should
  - Conduct a regular census of their postdocs;
  - Classify postdocs appropriately;
  - Establish appropriate, explicit policies for postdocs;
  - Adopt term limits for postdocs;
  - Provide a suitable post for postdocs to transition into if the term limit expires;
  - Periodically review the balance of interests of the postdoc, adviser, department, and institution to ensure adequate provisions are being made;
  - Not encourage unlimited growth of graduate and postdoctoral populations;
  - Maintain a postdoc office or officer;
  - Encourage the programs and divisions to assess their role in postdoctoral provision;
  - Require evidence that funding is available before allowing a postdoc hire;
  - Receive and keep on file a letter of postdoc appointment or contract;
  - Ensure that postdocs have career planning guidance;
  - Require a written evaluation from advisers annually on postdoc performance;
  - Address the needs of foreign postdocs;
  - Encourage and financially support a postdoctoral association.
- Funding agencies should
  - Work toward defining the postdoc;
  - Apply terms and conditions to all postdoc positions receiving funding;
  - Establish central offices and salary scales, and other measures, for postdocs;
  - Require a demonstration of the proper qualifications of those wishing to support postdocs.
- Scientific societies should
  - Play a larger role in promoting professional careers of postdocs;
  - Support postdoc job searches;
  - Develop norms for postdoc experiences in their field;
  - Collect and analyze data about career outcomes and employment prospects;
  - Organize programs or workshops to provide skills for postdocs;
  - Examine the purpose of the postdoc for their discipline.

Analysis

Overall, the report provides a very low bar for the administration and definition of the postdoc. While some of these recommendations have been largely achieved, it is disheartening to see how many basic recommendations about the postdoctoral position in this report remain to be implemented today.

Access to compensation, fringe benefits, institutional recognition, term limits, as well as the collection and provision of postdoc career data are still under discussion. Many institutions have term limits for postdocs, yet many either do not enforce them or simply change the title of the researcher. There are, however, institutions that take seriously the changes in title for postdocs and ensure that there is a transition to a permanent position with appropriate benefits. However, these changes can also start the tenure clock for those looking to transition into academic careers.

Many institutions set up policies for postdocs and maintain postdoctoral offices (see Chapter 3). This is largely due to the efforts of the National Postdoctoral Association, which was established in 2003 to advocate for improvement of the postdoctoral position. Their efforts have also led to formal definitions of postdoc titles used by the NIH and NSF. Annual reviews for postdocs are also becoming more common at universities, with individual development programs being a requirement for postdocs on NIH grant funding mechanisms.

Some of the actionable items that were discussed but not proposed in the report are also worth discussing. For example, the authors of the report were reluctant to suggest a reduction in the number of postdocs to funding agencies and institutions in 2000. In the intervening time period, however, the attitude within the academic community on this topic has largely shifted, and a consensus that there are too many postdocs in the current system has emerged. Disappointingly, there was also reluctance to set a formal benchmark for postdoc salaries in the 2000 report. The report suggests that recent evidence hints at a stabilization of the postdoctoral population, and thus the authors were loath to set fixed limits or measures on postdoc numbers and salaries. This conclusion turned out to be incorrect.

The postdoc recommendations in this report are sensible but limited by the availability of data. It is assumed that obtaining information on postdoc salaries, benefits, and institutional practices is possible. However, our own experience in searching for postdoc salary information in conjunction with the Fair Labor Standards Act (FLSA) revealed that most of this information is not freely available or easily accessible online. This issue is further complicated by the cultural perception that inquiring too much about financial gain is to imply some deficiency in the noble academic pursuit. Thus, data on postdocs are not provided because there are no expectations that they need to be by the scientific community.

Recommendations that advisers put expectations in writing have led to the creation of "compacts" signed between advisers and postdocs, such as that created by the Association of American Medical Colleges (AAMC, https://www.aamc.org/initiatives/research/postdoccompact/). There are some reports of advisers being unwilling to use such compacts due to the impression that by signing them, they may be used as legal documents. Also, with respect to providing career advice or assistance to postdocs, many willing advisers report that they are genuinely unable to do so—they are academics themselves and unlikely to know more about career options than their postdocs. This is where the institution has a responsibility to provide career resources and many postdocs (indeed, some graduate students) are currently unable to access the career resources or centers which their institutions provide for undergraduates or to access their institutional alumnae network. As a result, many postdocs (and even graduate students) have to create their own career resources at their institutions, in addition to the expectation that they create their own resources for other aspects of their training, such as in rigor and reproducibility (Baker, 2016).

Many institutions are still unable to report the exact number of employed postdocs, even in NSF surveys on the scientific workforce. There is an increasing need for an accurate census of postdoc researchers, and, as of yet, no report has provided a solution for helping institutions do so. Often this is due to the variety of titles for postdocs across institutions also identified as an issue 31 years prior to this report in "The Invisible University." Currently the American Society for Biochemistry and Molecular Biology and the National Postdoctoral Association are both leading efforts to harmonize postdoc titles at institutions that could directly address this barrier to data collection.

The Postdoctoral Experience Revisited (Committee to Review the State of Postdoctoral Experience in Scientists and Engineers et al., 2014)

Although the postdoctoral researchers themselves might feel invisible, there is broad recognition that something is amiss in the postdoctoral training system. Lack of data makes it difficult for leaders in research institutions and funding agencies to make policies about the role postdoctoral training should play in the research enterprise and for young people interested in science and engineering to make informed decisions on their career paths. Most postdoctoral researchers are employed by principal investigators to work on research grants, which creates an inherent source of stress. The investigator's primary mission is to complete the research, and any time spent in training the postdoctoral researcher is time not spent on the research. Principal investigators play an essential role in the training of postdoctoral researchers that must be acknowledged and reinforced by the funding agencies and institutions. Postdoctoral researchers are the future of the research enterprise, so it is critical that this period of training attract the most capable people to research.

The Postdoctoral Experience Revisited (Committee to Review the State of Postdoctoral Experience in Scientists and Engineers et al., 2014).

Overview

The title of the report deliberately references the 2000 report *"Enhancing the Postdoctoral Experience for Scientists and Engineers"* and aims to look at what has been achieved since that time in addition to what must still be achieved for postdocs. The report identifies the formation of the National Postdoctoral Association by postdocs themselves; the establishment of specific offices dedicated to postdocs at the NIH, NSF, and various institutions; and the creation of surveys and committees by other organizations as steps in the right direction. But it also highlights the fact that more tangible benefits such as data collection, salaries, mentoring, career resources/development and, indeed, visibility of postdocs have failed to be widely implemented. This report essentially brings us to the present-day state of postdoctoral affairs.

Recommendations

- Postdoctoral training should be for no longer than 5 years with allowances for family leave;
- Only those receiving advanced training in research should have a postdoc title;
- First-year graduate students should receive information about careers, particularly as postdoc positions are only meant for advanced training and not a default next step;
- The NIH should raise the NRSA level to $50,000 for Year 0 postdocs, and postdoc salaries should be set at this base level and adjusted for cost of living depending on the region;
- Institutions should provide postdocs with benefits appropriate for full-time employees;
- Host institutions should make provisions for postdocs to have multiple mentors and take responsibility for good mentoring;
- The NSF should serve as the primary collection point for postdoc data, including nonacademic and foreign researchers.

Analysis

This report, more than others discussed here, highlights the frustration at the lack of progress made in many of these areas, particularly in the recommendation for data collection: *"Recognizing that this recommendation on data collection has been made many times before with little effect, the committee stresses that research institutions and professional societies should explore what they can do to enrich what is known about postdoctoral researchers and that all institutions make better use of new technologies and social and professional networks to collect relevant and timely data."* The report makes pointed recommendations to particular stakeholders and attempts to make clear action-oriented directives rather than providing vague recommendations for improving the postdoc position.

The recommendations made in this report are yet to be implemented across academia. In particular, no known action has been taken to ensure that postdoctoral researchers are receiving training rather than simply substituting for permanent staff roles. Data on the training postdocs are receiving is extremely scarce and is limited to isolated programs, most significantly the NIH Broadening Experiences in Scientific Training Awards, which is limited to 17 institutions across the United States and funded for only a 5-year period (Feig et al., 2016; Layton et al., 2016; Mathur et al., 2015; Meyers et al., 2016). With regard to benefits, a recent NIH survey showed that most postdocs at surveyed institutions (the survey response rate being only 50%) received comparable health insurance benefits to other employees, but less annual leave, and access to retirement contributions, with smaller institutions appearing to treat postdocs better in these regards (Lauer, n.d.).

With regards to data collection on career outcomes of graduate students and postdocs, implementation is uneven, and efforts are most often focused on graduate students. Current proposed models for postdoc career outcome data collection include the recent analysis at the University of California, San Francisco (Silva et al., 2016). National efforts to promote consistent data collection for the purposes of a centralized dataset are currently being driven by the Rescuing Biomedical Research group.

SUMMARY

For brevity and to prevent repetition within the current report, a number of other published reports and articles have not only been described in detail but also been discussed the postdoctoral position (Anon, 1992, 2000; Biomedical Research Workforce Working Group, 2012; Coruzzi & Last, 1993; National Research Council (US) Committee on Bridges to Independence: Identifying Opportunities for and Challenges to Fostering the Independence of Young Investigators in the Life Sciences, 2005; National Research Council (US) Committee to Study the National Needs for Biomedical, Behavioral, and Clinical Research Personnel, 2011; Physician-Scientist Workforce Working Group, 2014; Reskin, n.d.; Zumeta, 1984). While there has been much discussion about the postdoctoral position over the last 50 years, it is striking how little progress has been made and what basic questions still remain unanswered, not least of which is an accurate estimate of the unknown postdoc population in the United States.

Why has there been such little progress given the many recommendations to improve the postdoc position over such a long period of time? To address this, we present an analysis that examines in detail the recommendations to increase postdoctoral salaries at the national level, and how the recent proposed updates to the FLSA affected them. This case study highlights key barriers, targets, and strategies for future efforts to implement postdoc reform in areas recently identified as requiring change (Pickett et al., 2015).

UPDATES TO THE FAIR LABOR STANDARDS ACT IN 2016

According to the US Department of Labor, the FLSA (United States Department of Labor, 2016a) "establishes minimum wage, overtime pay, recordkeeping, and youth employment standards affecting employees in the private sector and in Federal, State, and local governments."

On March 13, 2014, The White House issued a memorandum from US President Barack Obama (Obama, 2014) to the Secretary of Labor, Thomas Perez, to update the FLSA as follows:

> *I hereby direct you to propose revisions to modernize and streamline the existing overtime regulations. In doing so, you shall consider how the regulations could be revised to update existing protections consistent with the intent of the Act; address the changing nature of the workplace; and simplify the regulations to make them easier for both workers and businesses to understand and apply.*

On July 6, 2015, the Department of Labor issued a Notice of Proposed Rulemaking soliciting feedback by September 4, 2015. This notice proposed increasing the current FLSA exemption annual salary level from $23,660 (which was set in 2004) to $50,440 (which would be newly set in 2016), with automatic updates every 3 years (Obama, 2015; United States Department of Labor, 2016b). On May 18, 2016, the Secretary of Labor Thomas Perez updated the FLSA overtime regulations and in this the exemption, salary was lowered to $47,476 (United States Department of Labor, 2016e), with a set implementation date of December 1, 2016.

On September 28, 2016, the US House of Representatives passed a bill (HR 6094) by 246 votes to 177 to delay implementation of the FLSA ruling from the original December 1, 2016, date to June 1, 2017 (Congress of the United States of America, 2016). In preparation for the floor vote, a group of higher education organizations issued a letter to Congress in support of the bill arguing that the delay would allow universities and colleges more time to comply with the FLSA ruling (Anon n.d.-a, n.d.-b). A companion bill was also introduced by Senators James Lankford (R-OK), Lamar Alexander (R-TN), and Susan Collins (R-ME). The original bill was then passed to the Senate on September 29, 2016.

On October 12, 2016, 21 states (Nevada; Texas; Alabama; Arizona; Arkansas; Georgia; Indiana; Kansas; Louisiana; Nebraska; Ohio; Oklahoma; South Carolina; Utah; Wisconsin; Kentucky; Iowa; Maine; New Mexico; Mississippi; Michigan) filed an emergency motion for a preliminary injunction against the FLSA ruling (Anon, 2016), and on November 16, 2016, a preliminary injunction hearing took place to delay implementation of the FLSA ruling. At that point, if no delay were to be imposed, employers needed to comply with the FLSA on December 1, 2016. However, the injunction was granted nationwide on November 22, 2016 by a federal judge, thereby delaying implementation of the FLSA ruling. At the time of writing, the updates were still subject to the injunction, and their status not yet been clarified.

HOW THE FAIR LABOR STANDARDS ACT UPDATES INCLUDED POSTDOCS

Given the effects of the FLSA ruling on higher education, the College and University Professional Association for Human Resources (CUPA-HR) coordinated a letter to the Department of Labor on behalf of 18 Higher Education organizations with the following key recommendations: (1) longer time to adjust to the changes; (2) proposed lower salary level options of: $29,172, $30,004 or $40,352; and (3) rephrasing language to exempt postdocs based on their status as trainees, in a similar manner to medical residents (CUPA-HR, 2015a, 2015b; Brantley, 2016).

The Association of American Medical Colleges (AAMC) also submitted a letter to the Department of Labor similarly calling for "gradual and incremental" changes, an initial salary level below that of the NIH postdoctoral stipends ($42,840 in 2015), and an exemption for postdocs due to their status as "learned professionals" similar to medical residents (AAMC, 2015a, 2015b).

At the same time, postdocs individually submitted comments (Wexler, 2016) to the Department of Labor. On May 10, 2016, four unions representing postdocs or higher education employees (American Federation of State, County and Municipal Employees, Service Employees International Union, the United Auto Workers, and the National Education Association) met with the Department of Labor to argue against institutional calls for exemption (Penn, 2016).

On May 17, 2016, at the same time as the FLSA updates were released, it was also announced that postdocs would not be exempt from the FLSA ruling; in fact, they would be targeted by it (Collins & Perez, 2016). To be exempt, employees must pass all three of the following tests: the "salary basis test" (being paid on salary and not on an hourly basis), the "salary level test" (having a minimum salary of $913 per week or $47,476 per year, which does not apply to teachers), and the "standard duties test" (the employee's primary job duty had to pass an executive exemption, an administrative exemption, or a professional exemption). Postdocs fall under the professional exemption category. Since many postdocs also teach courses, the only exemption that could potentially apply to them was if their primary role is teaching. In addition, all international postdocs, regardless of visa status, as well as all postdocs on fellowships also were understood to be subject to the FLSA, as the key criterion to the FLSA is the nature of the work being done, not the immigration status or salary source of the employee, regardless of how the institution defines their postdocs as employees of "trainees."

DIRECT EFFECTS OF THE FAIR LABOR STANDARDS ACT ON POSTDOCS

The Department of Labor released a summary of the potential impact of the FLSA (United States Department of Labor, 2016d) and guidance documents for FLSA compliance (United States Department of Labor, 2016c) for higher education. The FLSA affects academics engaged primarily in research, including postdocs, whose status in terms of being trainees or employees has long been ambiguous. The Department of Labor classified all full-time postdocs whose primary role is not teaching as *"employees who conduct research at a higher education institution after the completion of their doctoral studies"* (United States Department of Labor, 2016c). Moreover, postdocs *"often meet the duties test for the* "learned professional" *exemption but must also satisfy the salary basis and salary level tests to qualify for this exemption"* (United States Department of Labor, 2016c).

Recommendations have been made consistently in academia to raise postdoctoral salaries (Pickett et al., 2015) to the level of at least $50,000, which is higher than the proposed level of $47,476 for overtime exemption. As a result of the FLSA updates, individual US institutions had the choice to increase the minimum postdoc salary to $47,476 or to classify postdocs as hourly workers and pay them overtime. The first option was problematic due to either individual faculty members having to come up with extra money to pay postdocs more or institutions having to provide bridge funding to support this. This scenario would also suggest that stipends for postdocs on fellowships would have to increase. On November 7, 2016, the NIH announced an increase to $47,484, just above the threshold of $47,476 (National Institutes of Health, 2016b). While NIH stipends are mandated by NRSA, it should be noted that most postdocs are funded from nontraining mechanisms and instead are funded by research project grants, which do not require that NRSA levels be followed; in addition, the NRSA levels are used often by institutions only as guidelines for their postdoc salary policies. Therefore, extra stipend sources would possibly only be forthcoming for postdoctoral researchers on fellowships or training mechanisms funded by NIH, such as F32 fellowship awards and T32 training grants.

Conversely, keeping track of overtime hours for postdocs would likely cost more in administrative burden (not least because these administrators themselves were also subject to the FLSA updates), in addition to running counter to academic culture, where the output of an academic's productivity is not measured by the number of hours they work. An additional burden was that to ensure salary raises for postdocs, all institutions now needed to identify all postdoctoral researchers to ensure compliance.

COMPLIANCE OF INSTITUTIONS WITH THE FAIR LABOR STANDARDS ACT RULING

To track how US institutions have complied with the FLSA ruling at the national level, we monitored university websites or contacted HR departments from universities across the country from the 2014 NSF Survey of GSS (National Science Foundation, n.d.-a, n.d.-b, n.d.-c), comprising a total of 340 institutions. The data were collected at the Future of Research online resource (http://futureofresearch.org/flsa-and-postdocs/) under the tab "Institutions and Funding Agencies: What they are doing" (Fig. 2.1). While not an accurate representation of the total postdoctoral population, the GSS numbers were used to provide our estimates of the percentage of the US postdoctoral population.

Our data as of October 21, 2016, comprising ~85% of the US postdoctoral population, showed that 41% of postdocs were expected to receive salary raises as of December 1, 2016 (see Fig. 2.2). We continued to collect data at various timepoints, and as of October 31, 2016, we had data for 97.5% of the US postdoctoral population (Bankston & McDowell, 2016). At that point, 51.2%

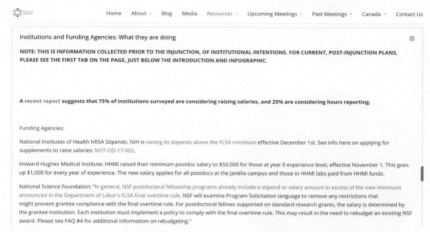

FIGURE 2.1 **Institutional compliance with the Fair Labor Standards Act (FLSA) ruling.** Screenshot of the Future of Research website with collected data focusing on how institutions complied with the FLSA ruling prior to the injunction. This information is found under the tab "Institutions and Funding Agencies: What they are doing" on the FLSA and postdocs website (http://futureofresearch.org/flsa-and-postdocs/).

FIGURE 2.2 **Postdoctoral population affected by Fair Labor Standards Act (FLSA) implementation plans: Oct 21, 2016.** Pie chart showing the percentage of the postdoctoral population at institutions implementing various plans for FLSA as of October 21, 2016.

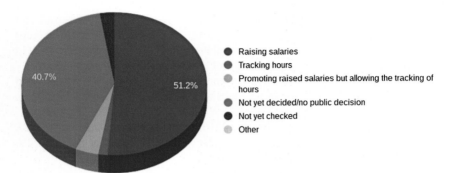

FIGURE 2.3   Postdoctoral population affected by Fair Labor Standards Act (FLSA) imple-mentation plans: Oct 31, 2016. Pie chart showing the percentage of the postdoctoral population at institutions implementing various plans for FLSA as of October 31, 2016.

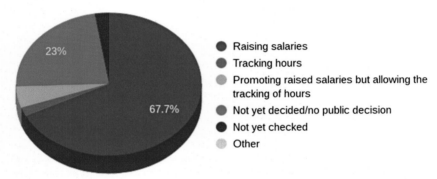

FIGURE 2.4   Postdoctoral population affected by the Fair Labor Standards Act (FLSA) implementation plans: Nov 10. Pie chart showing the percentage of the postdoctoral population at institutions implementing various plans for FLSA as of November 10, 2016.

of postdocs were expected to receive salary raises as of December 1, 2016 (Fig. 2.3). As of November 10, 2016, with 20 days to go, 67.7% of postdocs were expected to receive salary raises as of December 1, 2016 (Fig. 2.4). Finally, on November 21, 2016, with 10 days to go, 69.1% of postdocs were expected to receive salary raises as of December 1, 2016 (Fig. 2.5).

INSTITUTIONAL RESPONSE TO THE INJUNCTION AGAINST THE FAIR LABOR STANDARDS ACT UPDATES

On November 22, 2016, a federal court injunction blocked implementation of the FLSA updates on December 1, 2016. At this point, we began to track how institutions chose to respond to the injunction via checking the university web-sites or by contacting the HR departments of the same 340 institutions from

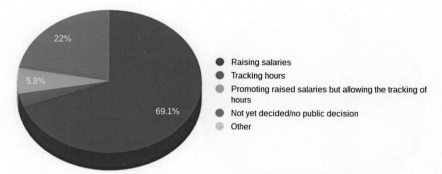

● Raising salaries
● Tracking hours
● Promoting raised salaries but allowing the tracking of hours
● Not yet decided/no public decision
○ Other

FIGURE 2.5   Postdoctoral population affected by the Fair Labor Standards Act (FLSA) implementation plans: Nov 20, 2016. Pie chart showing the percentage of the postdoctoral population at institutions implementing various plans for FLSA as of November 20, 2016.

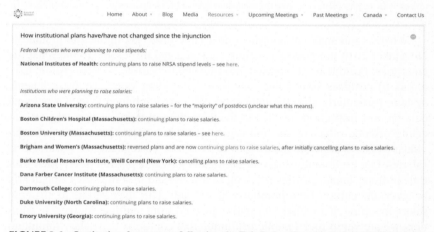

FIGURE 2.6   Institutional responses following the Fair Labor Standards Act (FLSA) injunction. Screenshot of the Future of Research website with collected data focusing on how institutions responded to the injunction against FLSA updates. This information is found under the tab "How institutional plans have/have not changed since the injunction" on the FLSA and postdocs website (http://futureofresearch.org/flsa-and-postdocs/).

the 2014 NSF Survey of Graduate Students and Postdoctorates in Science and Engineering (National Science Foundation, n.d.-a, n.d.-b, n.d.-c). The data are collected at the Future of Research online resource (http://futureofresearch.org/flsa-and-postdocs/) under the tab "How institutional plans have/have not changed since the injunction" (Fig. 2.6). On December 22, 2016, at exactly 1 month following the injunction, our data showed that 59.2% of postdocs were still expected to receive salary raises as of December 22, 2016 (Fig. 2.7), representing a decrease of 10%.

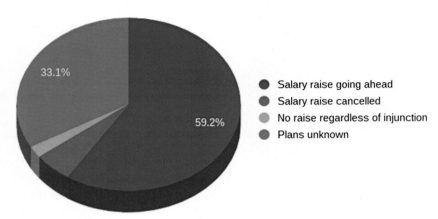

33.1%

59.2%

- Salary raise going ahead
- Salary raise cancelled
- No raise regardless of injunction
- Plans unknown

FIGURE 2.7 Postdoctoral population affected by the Fair Labor Standards Act (FLSA) injunction: Dec 22, 2016. Pie chart showing the percentage of the postdoctoral population at institutions with various plans following the injunction against FLSA updates as of December 22, 2016.

Whereas this may not appear to be a large drop, the injunction granted in November 2016 left postdocs feeling confused and disposable to the scientific enterprise. Postdoc reactions were documented by various sources (Anon, n.d.-a, n.d.-b) postdocs voiced that they were not feeling valued by the academic system, particularly after the injunction. We have continued to gather national data on how institutions reacted to the injunction in terms of pursuing or canceling postdoc salary raises, and began attempting to collect salaries of individual postdocs at public institutions to assess the current state of postdoctoral salaries. Our preliminary analysis shows that a sizable portion of reported full-time postdoc annual salaries reaching as low as (and in some cases, apparently beneath) the current legal minimum of $23,660. This may be due to errors in salary reporting, particularly if salaries are paid from multiple sources. In addition, we are yet to find a public institution that conforms to a salary minimum other than the legal minimum or that uses the NRSA stipend scale consistently, which is contrary to what might be assumed to be the current postdoctoral salary policy at US institutions. The difficulty that has been faced in attempting to get this most basic of information from institutions gives us cause for concern that the administration of postdocs at institutions is direr than currently supposed.

CONCLUSION

Postdocs have been at US institutions for at least 100 years, yet their status remains remarkably unchanged in that same time period. During this time, several reports have proposed reforms to improve the postdoctoral position. These reforms address a number of issues, including postdoc titles, transparency surrounding the postdoc position, work–life balance, and the instability of the postdoc position. There have also been several advocacy efforts for postdocs over the last several years. However, as evidenced from the still relatively static nature of the postdoctoral position during this time, such efforts are not sufficient to effect change. Organizations supporting or representing postdocs at the national level are now undertaking efforts to improve postdoc transparency by gathering data on several postdoc issues. An example is our most recent effort on the effects of the FLSA ruling and injunction on postdoctoral salaries. This particular case is useful to illustrate how gathering more data on postdoc issues could be used to effect change at individual institutions, with the goal of shifting the culture to allow for more transparency about the postdoc position in the future. In particular, it becomes possible to hold institutions accountable for what they say and do. Other efforts include the drive to harmonize postdoc titles led by the American Society for Biochemistry and Molecular Biology and the National Postdoctoral Association, and an effort to track career outcomes of graduate students and postdocs by Rescuing Biomedical Research.

With such a long history of calls for reform and so little action, what are the barriers to reform of the postdoctoral position? It appears not to be a question of ability, but of priority. It is simply not a priority of the academic system as a whole to ensure that postdocs are counted, trained, or granted a role other than providing the cheap labor required for research to be produced under a hypercompetitive system (Alberts et al., 2014). It is not clear that the academic community is truly making training of graduate students and postdocs a priority. Indeed, "The Postdoctoral Experience Revisited" shares this impression and states it clearly: "*any training that occurs is a byproduct of work*" (Committee to Review the State of Postdoctoral Experience in Scientists and Engineers et al., 2014). In assessing the steps taken to reform the postdoc position, a large proportion of the efforts mentioned in "The Postdoctoral Experience Revisited" appears to have been driven by postdocs themselves, from the creation of the National Postdoctoral Association and other grassroots organizations. Indeed, far from simply not enacting proposed reforms, the case of the updates to the FLSA demonstrates that institutions can actually act as a barrier to proposed changes, in their "successful" attempts to exempt postdocs from a ruling that would bring them closer, but not completely, in line with recent salary recommendations.

We advocate for recentering future actions by appreciating the current incentive structure of academia. By switching the discussion from "change is difficult" to "change is not desired," it is possible to see an increased role for junior researchers to drive change themselves, including involving the wider public. When most postdocs are "trainees" and federally funded, why is it not clear to the taxpayer how many there are, and where they go? What investment in training are postdocs themselves, and the public, making in this system? If change cannot be encouraged, it may be time to incentivize change by making data about the current system available. Increased transparency about academia will enable junior researchers to make informed choices about how they wish to pursue their careers and their research by finding career paths and research

environments that suit their needs. The data also hold institutions accountable. There was a clear "correct" answer to how to implement the updates to
the FLSA, by raising postdoctoral salaries. Postdocs and administrators alike
reported to us the effects that a publicly available list comparing institutions
had on encouraging change in the desired direction. Indeed, after the injunction against the updates on November 22nd, when some institutions canceled
raises in the days immediately afterward, by further tracking the response to
the injunction, we continued to receive updates in the subsequent months that
those decisions had again been reversed, and that salaries will be raised after all,
usually aiming to have raises in place by the 2018 financial year. Furthermore,
academia risks jeopardizing public investment in research by treating this funding in a cavalier attitude by populating the system with "trainees" but failing to
provide detailed evidence on how their training has been a worthwhile investment beyond measuring the number of papers produced and grant success rates.
Centering and prioritizing the development of future researchers, and not simply
relying on them to provide cheap labor, is a necessary transition for academia to
make if we are not to continue issuing similar recommendations, decade after
decade, without implementation.

ACKNOWLEDGMENTS

Many thanks to Lida Beninson, PhD, Yasmeen Hussain, PhD, Amanda Field, PhD, Irene
Ngun, PhD, and staff at the National Academies who have previously provided summaries
of key reports, which were of great help in verifying and checking recommendations about
postdocs. Thanks also to Chris Pickett, PhD and Alberto Roca, PhD for advice on reports and
articles in existence and general discussions.

REFERENCES

Association of American Medical Colleges. (2015a). *AAMC Submits Comments to Department of
Labor on Overtime Rule.*
Association of American Medical Colleges. (2015b). *AAMC Submits Comments to Department of
Labor on Overtime Rule. AAMC Submits Comments to Department of Labor on Overtime Rule.*
Available at https://www.aamc.org/advocacy/washhigh/highlights2015/443386/091815aamcsu
bmitscommentstodepartmentoflaboronovertimerule.html.
Alberts, B. (2010). Overbuilding research capacity. *Science, 329*(5997), 1257.
Alberts, B., et al. (2014). Rescuing US biomedical research from its systemic flaws. *Proc. Natl.
Acad. Sci. U.S.A., 111*(16), 5773–5777.
Anon. (1919). National research fellowships in physics and chemistry. *Proc. Natl. Acad. Sci., 5*(7),
313–315.
Anon. (1981). *Postdoctoral Appointments and Disappointments.* Washington, DC.: National
Academies Press.
Anon. (1992). *Educating Mathematical Scientists: Doctoral Study and the Postdoctoral Experience
in the United States.* Washington, DC: National Academies Press.
Anon. (2000). *Addressing the Nation's Changing Needs for Biomedical and Behavioral Scientists.*
Washington, DC: National Academies Press.

Anon. (2016). *State of Nevada et al v. United States Department of Labor, et al, Docket No. 4:16-cv-00731. State of Nevada et al v. United States Department of Labor et al, Docket No. 4:16-cv-00731.* Available at https://www.bloomberglaw.com/public/desktop/document/State_of_Nevada_et_al_v_United_States_Department_of_Labor_et_al_D/1?1476407516.

Anon. n.d.-a. US Postdocs Grapple with Salary Changes | The Scientist Magazine®. US Postdocs Grapple with Salary Changes | The Scientist Magazine®. Available at: http://www.the-scientist.com/?articles.view/articleNo/47800/title/US-Postdocs-Grapple-with-Salary-Changes/.

Anon. n.d.-b. www.acenet.edu/news-room/Documents/Letter-ED-Workforce-Committee-Overtime-Rule-Delay.pdf.

Baker, M. (2016). Reproducibility: seek out stronger science. *Nature, 537*(7622), 703–704.

Bankston, A., & McDowell, G. S. (2016). Monitoring the compliance of the academic enterprise with the fair labor standards act. [version 1; referees: 2 approved, 1 approved with reservations] *F1000Research, 5,* 2690.

Berelson, B. (1960). *Graduate Education in the United States.* McGraw-Hill Book Company. Incorporated. Available at https://books.google.com/books?id=wgwJAQAAIAAJ.

Berelson, B. (1962). Postdoctoral work in american universities: a recent survey. *J. High. Educ., 33*(3), 119.

Biomedical Research Workforce Working Group. (2012). *Biomedical Research Workforce Working Group Report. (Report to the Advisory Committee to the Director).* Available at http://acd.od.nih.gov/Biomedical_research_wgreport.pdf.

Bourne, H. R. (2013). A fair deal for PhD students and postdocs. *eLife, 2,* e01139.

Brantley, A. (2016). *Proposed Overtime Changes: Too Much for Higher Ed.* The Hill. Available at http://thehill.com/blogs/congress-blog/economy-budget/278388-proposed-overtime-changes-too-much-for-higher-ed.

Collins, F. S., & Perez, T. E. (2016). *Fair Pay for Postdocs: Why We Support New Federal Overtime Rules.* Huffington Post. Available at http://www.huffingtonpost.com/francis-s-collins-md-phd/fair-pay-for-postdocs-why_b_10011066.html.

Committee to Review the State of Postdoctoral Experience in Scientists and Engineers., et al. (2014). *The Postdoctoral Experience Revisited.* Washington, DC: National Academies Press (US).

Congress of the United States of America. (2016). *H.R.6094–114th Congress (2015-2016): Regulatory Relief for Small Businesses, Schools, and Nonprofits Act.* Available at https://www.congress.gov/bill/114th-congress/house-bill/6094/actions?q=%7B.

Coruzzi, G. M., & Last, R. L. (1993). A crucial role for the NSF postdoctoral fellowship program in plant biology. *Plant Cell, 5*(7), 722.

CUPA-HR. (2015a). *DOL's Proposed Overtime Rule: Higher Education's Comments & Concerns.* Available at http://www.cupahr.org/advocacy/files/FLSA_OT_Talking_Points_Sept2015.pdf.

CUPA-HR. (2015b). *Re: Notice of Proposed Rulemaking; Defining and Delimiting the Exemption for Executive, Administrative, Professional, Outside Sales and Computer Employees (80 Fed. Reg. 38515, July 6, 2015) (RIN 1235–AA11).* Available at http://www.naicu.edu/docLib/20150904_CUPA-HR_FLSA_comment_letter.pdf.

Curtis, R. B. (1969). *Invisible University: Postdoctoral Education in the United States. Report of a Study Conducted under the Auspices of the National Research Council.* Washington, DC: National Academies Press.

Faupel-Badger, J. M., Nelson, D. E., & Izmirlian, G. (2017). Career satisfaction and perceived salary competitiveness among individuals who completed postdoctoral research training in cancer prevention. *PLoS One, 12*(1), e0169859.

Feig, A. L., et al. (2016). Using longitudinal data on career outcomes to promote improvements and diversity in graduate education. *Change, 48*(6), 42–49.

Ferguson, K., et al. (2014). *National Postdoctoral Association Institutional Policy Report 2014: Supporting and Developing Postdoctoral Scholars*. National Postdoctoral Association. Available at http://www.nationalpostdoc.org/?policy_report_databa.

Franzoni, C., Scellato, G., & Stephan, P. (2015). International mobility of research scientists. In *Global Mobility of Research Scientists* (pp. 35–65). Elsevier.

Fuhrmann, C. N., et al. (2011). Improving graduate education to support a branching career pipeline: recommendations based on a survey of doctoral students in the basic biomedical sciences. *CBE Life Sci. Educ.*, *10*(3), 239–249.

Garrison, H. H., Justement, L. B., & Gerbi, S. A. (2016). Biomedical science postdocs: an end to the era of expansion. *FASEB J.*, *30*(1), 41–44.

Heggeness, M. L., et al. (2016). *The U.S. Biomedical Workforce: Using Historical Decennial Census Data to Inform Policies, Programs, and Career Decision-Making*.

Heggeness, M. L., et al. (2017). The new face of US science. *Nature*, *541*(7635), 21–23.

Israeli, N. (1944). American postdoctoral education. *J. High. Educ.*, *15*(8), 428.

Johns Hopkins University. (1914). *Graduates and Fellows of the Johns Hopkins University, 1876–1913*. Baltimore: Johns Hopkins Press.

Kahn, S., & Ginther, D. K. (2017). The impact of postdoctoral training on early careers in biomedicine. *Nat. Biotechnol.*, *35*(1), 90–94.

Lauer, M., National Institutes of Health Office of Extramural Research. Update on the Postdoctoral Benefit Survey. Available at: https://nexus.od.nih.gov/all/2015/11/30/update-postdoctoral-benefit-survey/.

Layton, R. L., et al. (2016). Diversity exiting the ccademy: influential factors for the career choice of well-represented and underrepresented minority scientists. *CBE Life Sci. Educ.*, *15*(3).

Mathur, A., et al. (2015). Transforming training to reflect the workforce. *Sci. Transl. Med.*, *7*(285), 285ed4.

McDowell, G. S. (2016). *ASBMB Today. Four Reasons We Don't Need 37 Names for Postdocs*. Available at http://www.asbmb.org/asbmbtoday/201604/Education/Postdoc/.

Meyers, F. J., et al. (2016). The origin and implementation of the Broadening Experiences in Scientific Training programs: an NIH common fund initiative. *FASEB J.*, *30*(2), 507–514.

National Academy of Sciences, National Academy of Engineering, Institute of Medicine, & Committee on Science, Engineering, and Public Policy. (2000). *Enhancing the Postdoctoral Experience for Scientists and Engineers*. Washington, DC: National Academies Press.

National Institutes of Health. Kirschstein-NRSA Stipend History. Kirschstein-NRSA Stipend History. Available at: http://researchtraining.nih.gov/sites/default/files/docs/Stipend_History_V2.xls.

National Institutes of Health. (2016a). *NIH Almanac: Chronology of Events. NIH Almanac: Chronology of Events*. Available at http://www.nih.gov/about/almanac/historical/chronology_of_events.htm.

National Institutes of Health. (2016b). *NOT-OD-16–134: Revised: Projected FY 2017 Stipend Levels for Postdoctoral Trainees and Fellows on Ruth L. Kirschstein National Research Service Awards (NRSA). NOT-OD-16-134: Revised: Projected FY 2017 Stipend Levels for Postdoctoral Trainees and Fellows on Ruth L. Kirschstein National Research Service Awards (NRSA)*. Available at https://grants.nih.gov/grants/guide/notice-files/NOT-OD-16-134.html.

National Postdoctoral Association. NPA Recommendations for Postdoctoral Policies and Practices - National Postdoctoral Association. NPA Recommendations for Postdoctoral Policies and Practices - National Postdoctoral Association. Available at: http://www.nationalpostdoc.org/?recommpostdocpolicy.

National Research Council (US) Committee on National Needs for Biomedical and Behavioral Research Personnel. (1994). Meeting the Nation's needs for biomedical and behavioral scientists. In *The National Academies Collection: Reports Funded by National Institutes of Health*. Washington, DC: National Academies Press (US).

National Research Council (US) Committee on Dimensions, Causes, and Implications of Recent Trends in the Careers of Life Scientists. (1998). Trends in the early careers of life scientists. In *The National Academies Collection: Reports Funded by National Institutes of Health*. Washington, DC: National Academies Press (US).

National Research Council (US) Committee on Bridges to Independence: Identifying Opportunities for and Challenges to Fostering the Independence of Young Investigators in the Life Sciences. (2005). Bridges to independence: fostering the independence of new investigators in biomedical research. In *The National Academies Collection: Reports Funded by National Institutes of Health*. Washington, DC: National Academies Press (US).

National Research Council (US) Committee to Study the National Needs for Biomedical, Behavioral, and Clinical Research Personnel. (2011). Research training in the biomedical, behavioral, and clinical research sciences. In *The National Academies Collection: Reports Funded by National Institutes of Health*. Washington, DC: National Academies Press (US).

National Science Foundation. (2017). *Postdoctoral Research Fellowships in Biology (PRFB)*. Available at https://www.nsf.gov/publications/pub_summ.jsp?WT.z_pims_id=503622.

National Science Foundation. n.d.-a. Early Career Doctorates Survey (ECDS). Early Career Doctorates Survey (ECDS). Available at: https://www.nsf.gov/statistics/srvyecd/.

National Science Foundation. n.d.-b. Survey of Graduate Students and Postdoctorates in Science and Engineering (GSS). Available at: http://www.nsf.gov/statistics/srvygradpostdoc/#tabs-1.

National Science Foundation. n.d.-c. Survey of Doctorate Recipients (SDR). Available at: http://www.nsf.gov/statistics/srvydoctoratework/.

Obama, B. (2014). *Presidential Memorandum – Updating and Modernizing Overtime Regulations*. Available at https://www.whitehouse.gov/the-press-office/2014/03/13/presidential-memorandum-updating-and-modernizing-overtime-regulations.

Obama, B. (2015). *A Hard Day's Work Deserves a Fair Day's Pay. Huffington Post*. Available at http://www.huffingtonpost.com/barack-obama/a-hard-days-work-deserves-a-fair-days-pay_b_7691922.html.

Penn, B. (2016). *Employers to White House: Give Us More Time on Overtime Rule. Bloomberg*. Available at http://www.bna.com/employers-white-house-n57982072491/.

Physician-Scientist Workforce Working Group. (2014). *Physician-Scientist Workforce Working Group Report*. Available at http://f1000.com/work/item/2799447/resources/2088225/pdf.

Pickett, C. L., et al. (2015). Toward a sustainable biomedical research enterprise: finding consensus and implementing recommendations. *Proc. Natl. Acad. Sci. U.S.A.*, *112*(35), 10832–10836.

Polka, J. (2014). *American Society for Cell Biology COMPASS Blog Where Will A Biology PhD Take You?*. Available at http://www.ascb.org/where-will-a-biology-phd-take-you/.

Polka, J. K., Krukenberg, K. A., & McDowell, G. S. (2015). A call for transparency in tracking student and postdoc career outcomes. *Mol. Biol. Cell*, *26*(8), 1413–1415.

Powell, K. (2012). The postdoc experience: high expectations, grounded in reality. *Science*, 992–996.

Reskin, B. F. (1976). Sex differences in status attainment in science: the case of the postdoctoral fellowship. *Am. Sociol. Rev.*, *41*(4), 597–612. Available at http://www.jstor.org/stable/2094838.

Silva, E. A., et al. (2016). Tracking career outcomes for postdoctoral scholars: a call to action. *PLoS Biol.*, *14*(5), e1002458.

Stephan, P. E., & Levin, S. G. (2001). Exceptional contributions to US science by the foreign-born and foreign-educated. *Popul. Res. Policy Rev.* Available at http://link.springer.com/article/10.1023/A:1010682017950.

United States Department of Labor. (2016a). *Compliance Assistance - Wages and the Fair Labor Standards Act (FLSA) - Wage and Hour Division (WHD)*. Available at https://www.dol.gov/whd/flsa/.

United States Department of Labor. (2016b). *Defining and Delimiting the Exemptions for Executive, Administrative, Professional, Outside Sales and Computer Employees*. Available at http://webapps.dol.gov/FederalRegister/HtmlDisplay.aspx?DocId=28355&AgencyId=14.

United States Department of Labor. (2016c). *Guidance for Higher Education Institutions on Paying Overtime under the Fair Labor Standards Act*.

United States Department of Labor. (2016d). *Overtime Final Rule and Higher Education*. Available at https://www.dol.gov/sites/default/files/overtime-highereducation.pdf.

United States Department of Labor. (2016e). *The Overtime Rule*. Available at https://www.dol.gov/featured/overtime/.

Wexler, E. (2016). What does the Department of Labor's overtime rule mean for higher education? *Inside High. Ed.* Available at https://www.insidehighered.com/news/2016/02/25/what-does-department-labors-overtime-rule-mean-higher-education.

Zumeta, W. (1984). Anatomy of the boom in postdoctoral appointments during the 1970s: troubling Implications for quality science? *Sci. Technol. Hum. Values*, 9(2), 23–37. Available at http://www.jstor.org/stable/689540.

Chapter 3

Institutional Support, Programs, and Policies for Postdoctoral Training

Nisha A. Cavanaugh
Sanford Burnham Prebys Medical Discovery Institute (SBP), La Jolla, CA, United States

Chapter Outline

Addressing the Employment
Challenges of Postdocs   51
Defining the Postdoctoral Training
Period and Implementing Individual
Development Plans   56
Career and Professional
Development Programming   58

Alumni Databases: Tracking Postdoc
Career Outcomes   63
Final Thoughts   64
Acknowledgments   65
References   66

This chapter focuses on some of the multifaceted issues (beyond research alone) facing postdocs and their support system. It is intended to help all involved stakeholders to appreciate the challenges of effective mentoring and the many perspectives from which postdoctoral training can be viewed. Special attention is paid to the institutional postdoctoral office, which provides a support structure and framework to effectively engage all participants. Many of the topics discussed below may be of particular use to young mentors and institutions without an official postdoc office that are unaware of the abundant, readily available resources to enhance the postdoctoral experience.

The purpose of a postdoctoral (postdoc) position was originally to provide a temporary period of mentored training for a PhD to develop into an independent scientist and then transition into an academic tenure-track faculty position. As a highly educated labor force, postdocs are invaluable to Principal Investigators (PIs); they are often the primary drivers of academic research through publishing papers, submitting grant applications for funding, managing the laboratory, and training or supervising junior laboratory members. In exchange, postdocs are compensated with a higher salary than graduate students and receive mentoring from their PI as they establish themselves as independent researchers.

The Postdoc Landscape. http://dx.doi.org/10.1016/B978-0-12-813169-5.00003-3

In the 1970s, only 21% of doctorates spent more than 4 years in postdoctoral positions (National Research Council, 2005). Over time, that number has steadily increased due to a combination of factors including an increased number of PhDs being awarded and a limited number of available bench research positions (in academia, industry, or government) (Kahn & Ginther, 2017; Mason et al., 2016). Postdocs have unique challenges since they are considered to have high-level analytical, technical, and research skills (acquired by completing a doctoral program), and yet they are viewed by faculty and other established researchers as needing additional training toward obtaining independent positions. Often, postdocs are deemed temporary (no longer students and not considered employees) that require more training and development before entering the job market. Over the course of their training period, postdocs are expected to demonstrate research independence and move on to more permanent positions. At the same time, the majority of postdoctoral scholars are at the stage in their lives (predominantly in their 30s) (Davis, 2005) where they have a young family or are considering starting one, which means they are (or will be) in a financial situation to provide for that family and start saving funds for the future.

As the number of postdocs has increased over the last 2 decades and the feasibility of securing a research position has decreased, a multifaceted and multiorganizational approach has been implemented to change targeted aspects of the postdoctoral experience (Committee to Review). While government agencies (particularly the National Science Foundation (NSF) and the National Institutes of Health (NIH)) and the National Postdoctoral Association (NPA) discuss postdoctoral issues on a national scale, research institutions and universities have established postdoctoral offices and administrative units to address the specific needs of their local postdoctoral populations and enhance the overall training experience. In the early 2000s, there were fewer than 25 offices committed to serving postdoctoral scholars; according to the NPA's recent Institutional Policy Report, in 2014 this number grew to at least 167 postdoctoral offices nationwide serving the needs of approximately 79,000 postdoctoral scholars (Ferguson, Huang, Beckman, & Sinche, 2014).[1] Postdoctoral administrative units most commonly report to an institution's Graduate School; however, many postdoc offices can be located in different areas or supervised by various offices:

- within the Research Affairs office,
- report to an Academic/Administrative Dean or Provost/Chancellor/President,
- or report to an institution's Human Resources or Training department (Ferguson et al., 2014).

---

1. The National Postdoctoral Association Institutional Report reflects data from member offices or units; additional offices may exist.

The mission of postdoctoral offices is to work with all institutional stake-holders (including faculty, postdocs, and executive leadership) and enhance the postdoctoral training experience in several ways:

- serving as the centralized resource for institutional recommendations and policies that pertain to the employment and/or training of postdocs,
- providing professional development programs (which complement research training) to help postdocs develop and hone skills, toward enhancing their success as developing scientists,
- providing career development programs to help postdocs explore various career paths, network with professionals, and prepare for the job market, and
- developing and maintaining a mechanism to track where postdocs go after they leave the institution and how their training experience may have influenced their career outcome.

Based on my 8 years of experience (which includes working as a postdoctoral scholar, volunteering with the NPA Advocacy Committee, and serving as a postdoc office administrator), I have identified four distinct areas in which postdoc offices and institutions currently have the greatest impact on postdocs: (1) addressing the challenges of postdoctoral appointments in terms of salary and benefits, (2) defining the postdoctoral training period and implementing Individual Development Plans (IDPs), (3) providing career and professional development programs, and (4) tracking and maintaining alumni databases. These areas fall primarily under the purview of postdoc offices because they are uniquely positioned to uphold and maintain institutional policies and recommendations while understanding and appreciating the institutional culture and climate, and also to recognize the unique perspective of their postdocs. For example, a research institution may only have postdocs specializing in biomedical research, while a state university will have postdocs in a wide range of disciplines (from engineering to social sciences); these two institutions will have postdoc offices serving postdoc populations with very different needs. While there are other topics affecting postdocs that can be discussed, I have found these four areas are common to all institutions and demonstrate how postdoctoral offices continue to have the greatest impact on enhancing the quality of the postdoctoral training experience.

ADDRESSING THE EMPLOYMENT CHALLENGES OF POSTDOCS

It has been well-documented that postdocs are undervalued monetarily (Kahn & Ginther, 2017; National Research Council, 2000). Postdocs earn comparable salaries to an average US worker, yet they work nearly 20h more per week (Stephan, 2013). This discrepancy highlights the significant financial investment that postdocs must make into education rather than entering the workforce, since they may be missing out on valuable years of on-the-job training. According to economist Paula Stephan, scientists that graduated with a Bachelor's degree but did not pursue a PhD earn $20,000 a year more and possibly receive better benefits than their counterparts who completed a doctoral degree program (Stephan, 2013). A recent study compared salaries over 10 years after participants received doctoral degrees and found that PhDs (with and without postdocs) receive $12,000 less on average in salary than those who do not have this advanced degree (Kahn & Ginther, 2017). For example, the median starting salary for a postdoc is $44,724, while PhDs who go directly into industry after graduate school earn $73,662 (Kahn & Ginther, 2017). Ironically, many postdocs will still only receive entry-level pay on entering industry despite years of additional research training. The issue is further compounded by the fact that many of these individuals may be at a stage of great financial need (supporting a young family or taking care of elderly relatives); these postdocs may find themselves overworked and undercompensated.

Until recently, there were no federal regulations regarding postdoc salary except for those postdocs who received the NIH National Research Service Award (NRSA) fellowships or in cases where institutions required postdoc salaries to follow the NRSA stipend minimum of $43,692 (Table 3.1) (Revised: Projected FY, 2016). However, not all postdocs are supported by NIH, and many other federal agencies and private foundations do not have clearly expressed salary minimums. As a result, minimum postdoc salaries vary greatly between institutions. According to the 2014 NPA Institutional Policy Report, the range of minimum salaries for postdocs reported by its institutional members varied from mid-$20K to $80K (Ferguson et al., 2014).

TABLE 3.1 Projected FY 2017 Stipend Levels for Postdoctoral Trainees and Fellows on Ruth L. Kirschstein National Research Service Awards (NRSA) (Revised: Projected FY, 2016)

Years of Postdoctoral Experience	Actual Stipend for FY 2016	Projected Stipend for FY 2017
0	$43,692	$47,484
1	$45,444	$47,844
2	$47,268	$48,216
3	$49,152	$50,316
4	$51,120	$52,140
5	$53,160	$54,228
6	$55,296	$56,400
7 or more	$57,504	$58,560

The stipend levels described in this table reflect how much postdocs were paid before Fair Labors Standard Act (FLSA) (FY 2016) and after FLSA (FY 2017). Stipend levels increase with years of postdoctoral experience, ensuring that postdocs who are supported on National Institutes of Health NRSA grants will receive annual salary increases.

More recently, the postdoc salary minimum was addressed by updates to the Fair Labors Standard Act (FLSA).

The changes to FLSA ensure that all postdocs (who are not in a primarily teaching role), regardless of visa or fellowship status, would be compensated at a minimum salary of $47,476. As discussed in Chapter 2, many institutions have already made plans to implement the new minimum salary under FLSA for their postdoctoral scholars and PhDs in nonfaculty positions. In response to the proposed FLSA salary minimum, NIH established new NRSA stipend minimums that are commensurate with years of postdoctoral experience (see Table 3.1; note the stipend minimum at year=0 is a little higher at $47,484 compared to the FLSA minimum). Postdocs who are supported on NRSA fellowships automatically receive pay increases as reflected by the NIH pay scale (Table 3.1). This aligns with a report from the National Research Council that suggests raising salaries of postdocs to "appropriately reflect their value and contribution to research,"(Committee to Review; National Research Council, 2000) a suggestion that supports the recommendation of the NIH Biomedical Workforce Group to increase postdoctoral stipends, recognizing that the long training period in combination with a low salary can make a career in the biomedical sciences less attractive than other scientific disciplines and professional careers (NIH Biomedical Research Workforce).

FLSA and the updated NIH NRSA postdoc salary minimums begin to address the problem outlined by Paula Stephan and others. However, it is a complex problem since salary support typically comes from the faculty mentor (unless the postdoc is supported on an institutional training grant or on his/her own fellowship). Since PIs have not been given an automatic increase in existing funds to cover the new NIH salary minimum, they must stretch their existing grant money or request institutional support from overhead or discretionary funds. This can further impact available resources to employ postdocs for multiple years or affect recruitment to bring in new postdocs. This transition has shown the complexity and difficulty involved in raising postdoctoral compensation across the board.

The role of the institutions and postdoc offices is to provide support and guidance to postdocs and faculty as well as information about salary scales (if they exist). Additionally, to provide a mechanism for postdocs to be considered for salary increases incrementally over time, many institutions encourage an annual review process wherein postdocs are evaluated on their performance (similar to performance evaluation for permanent employees). Further discussion is needed to determine the metrics by which a postdoc's productivity or performance can be measured; nonetheless, an annual evaluation is an opportunity for discussion and communication of expectations between the postdoc and the PI/mentor. Such a review provides one route to a mutually beneficial, merit-based environment.

Another benefit that some institutions offer exceptional candidates are postdoctoral fellowships in which salaries are higher than the NIH NRSA postdoc salary minimums. For example, Sanford Burnham Prebys Medical Discovery Institute in California has created the Fishman Fund Fellowship in recognition of the institution's founders, Dr. William and Lillian Fishman (Fishman Fund Fellowship at SBP). This 2-year fellowship program is awarded to one exceptional postdoc who demonstrates a clear research proposal/plan. The postdoc receives a minimum $60,000 per year stipend that increases in the second year in addition to funds to pursue career development opportunities. Similarly, the Beckman Institute in Illinois has a Postdoctoral Fellowship, funded by the Arnold and Mabel Beckman Foundation, which provides a $54,000 annual stipend, benefits, and a research budget (Beckman Institute Postdoctoral Fellows Program). These mechanisms are attractive not only for the higher salary but also because they are open to all postdocs regardless of visa status, unlike some federally funded fellowship programs. Postdocs who successfully obtain a fellowship through their institution also have evidence of obtaining funding, and they can add this accomplishment to their curriculum vitae or resume when they are on the job market.

Salary is only one side of compensation; benefits packages (which may include health, dental, vision, leave policies, and retirement) can also vary widely among institutions and within a single institution depending on the funding mechanism. Postdocs are often supported through three different types of mechanisms: institutional training grants, individual fellowships (through the postdoc's home country, federal government agency, or a private foundation), and faculty-obtained research grants, with the latter being most common. Unlike most full-time employees, postdoc employment does not guarantee access to benefits; the mechanism by which postdocs are paid can influence their employment classification and thus the benefits package for which they are eligible. According to a National Council of University Research Administrators survey (on a recommendation by the Advisory Committee to the NIH Director working group), 91% of postdocs from 167 institutions received health insurance and 88% of postdocs received dental insurance (Lauer, 2015). This survey also indicated that institutions with smaller numbers of postdocs were more likely to receive health and/or retirement benefits compared to institutions with larger number of postdocs, while 44 institutions reported none of their postdocs receive retirement benefits (Lauer, 2015). The 2014 NPA Institutional Policy Report showed that postdocs supported on their PI's research grants had the highest likelihood to have health, dental, and/or vision benefits for themselves and/or their families (95% receiving health insurance, 91% dental insurance, and 80% vision insurance), while individuals on fellowships funded by their home countries were the least likely to have insurance for themselves and/or their families (Ferguson et al., 2014).

The NPA Institutional Survey also indicated that most institutions offered postdocs unpaid time off and family leave, and only a handful addressed paid family leave policies for postdocs (Ferguson et al., 2014). Moreover, few institutions offered services to support young families, such as on-site childcare facilities or subsidized childcare costs. This was recently documented in a study that examined the experiences of postdoc parents. The final report identified areas, such as unpaid leave and scarce parental resources, having an impact on postdoc retention (Lee, Williams, & Li, 2017). While postdocs may be temporary, they are at a stage in their lives (the majority are in their late 20 and 30s) (Davis, 2005) in which they often have young children and would greatly benefit from paid parental and family leave policies as well as optional retirement plans. Institutions, professional societies, and funding agencies continue to have conversations centered on possibilities for postdoc benefits.

Postdoc offices can serve as a resource to inform all postdocs and faculty about Title IX (a federal law) and leave policies. According to Title IX: *No person in the United States shall, on the basis of sex, be excluded from participation in, be denied the benefits of, or be subjected to discrimination under any education program or activity receiving Federal financial assistance* (The Pregnant Scholar, Title IX Basics). For the purposes of Title IX, trainee postdocs can be characterized as "students" or "anyone who is getting an education (even hands-on education) as a part of a covered program" (The Pregnant Scholar, What Laws Protect Postdocs?). This means that they are entitled to certain benefits including parental leave on the addition of a family member (through birth or adoption) (The Pregnant Scholar, Funding Parental Leave). While many postdocs may have job security—meaning they would not lose their job while on parental leave—they are not necessarily ensured pay under Title IX and will essentially be on unpaid leave.

Furthermore, depending on the institution and how the postdocs are funded, some postdocs may not accrue vacation and/or sick leave, which presents a challenge for postdoc parents any time a child requires care (e.g., due to an illness or a scheduled doctor's appointment). For example, externally funded postdocs (e.g., individuals supported by a fellowship from their home country) may miss out on certain benefits since it may not be clear who is responsible for providing those benefits (the institution or the funding agency). As a result, the lack of financial support and benefits for PhDs and postdocs may impact the workforce and recruitment of postdocs both within the United States and internationally; while research in the United States continues to have a strong reputation, employees' benefits are viewed as inferior compared to other countries (Committee to Review; Pew Research Center).

What can institutions do? The postdoc office serves as a centralized resource for disseminating employment and benefit information to postdocs and their faculty mentors so that they are aware of the relevant policies and recommendations that pertain to postdocs. This is often achieved through orientation sessions for postdocs (and information sessions for faculty) in which postdoc offices partner with human resources departments to present the benefits and retirement (if it exists) package options available to postdocs. Institutions can also keep postdocs more informed about their options as circumstances change (e.g., addition of a child or marriage) by posting information clearly on institutional websites and offering information sessions on a regular basis. Lastly, postdoc offices can bring in financial advisors or speakers to talk about managing budgets and how to make financial decisions based on limited budgets.

Although postdoc offices do not make policies themselves, they are uniquely positioned to bring together the various stakeholders—executive leaders, faculty, and postdocs—to discuss relevant issues facing postdocs and make recommendations if action or policy can be made at the institutional level. For example, postdoc offices can gather data on the number of postdocs that are parents within the institution, and utilize exit surveys to track in what ways (if any) current policies and benefits impact a postdoc's training experience and career trajectory. In the next section, I will discuss how institutions can define and limit the postdoctoral training period to facilitate the transition of postdocs into nontraining positions that include retirement, leave, and access to employee benefits.

DEFINING THE POSTDOCTORAL TRAINING PERIOD AND IMPLEMENTING INDIVIDUAL DEVELOPMENT PLANS

As noted, the NIH and the NSF define a postdoc as: *An individual who has received a doctoral degree (or equivalent) and is engaged in a temporary and defined period of mentored advanced training to enhance the professional skills and research independence needed to pursue his or her chosen career path* (Letter from National Science Foundation's Office). The supply of graduate students and postdoctoral scholars significantly outpaces the total number of available jobs for PhDs (Schillebeeckx, Maricque, & Lewis, 2013). In 2014, the difference between the supply of PhDs and the demand of available jobs was nearly 10,000 (Mason et al., 2016). While only 21% of postdoc positions lasted more than 4 years in the 1970s (National Research Council, 2005), since then, more positions have increased in duration and some individuals have opted to accept a second postdoc position after completing the first. As a result, individuals who completed a doctoral degree program and engaged in postdoctoral training may take 10–15 years before starting a more permanent position (i.e., a real job) (Rockey, 2012).

To determine how long postdocs are staying in training types of positions, institutions are finding ways to clearly classify and track their postdocs with standard job titles that are consistent across the country. Most postdocs are classified as "Postdoctoral Associate, Postdoctoral (Teaching or Research) Scholar, and/or Postdoctoral Fellow"; however, some institutions also consider "Research Associate" and "Research Scholar," among other similar titles, comparable to postdoc positions. Tracking this information will not only help determine how long PhDs stay in postdoc positions, but the data will also help with national surveys (e.g., Survey of Earned Doctorates or the Survey of Graduate Students and Postdoctorates in Science and Engineering) and determining career outcomes (see Section Alumni Databases: Tracking Postdoc Career Outcomes of this chapter).

Based on the NIH's recommendation that federal funding for National Research Service Awards and research grants should not exceed 5 years for postdoctoral training (National Research Council, 2005), many institutions have adopted term limits and capped the postdoctoral training period at 5 years (63% according to the 2014 NPA Institutional Survey) (Ferguson et al., 2014). However, this term limit may not necessarily include previous years of postdoctoral training at other institutions (e.g., the individual is engaging in a second postdoc). To ensure the value of training, when a postdoc stays at an institution for more than 5 years, promotion options should be explored in which a postdoc could potentially be elevated to a staff scientist or equivalent position (Alberts, Kirschner, Tilghman, & Varmus, 2014; Committee to Review). Elevating the employment status of a postdoc would not only ensure that the individual has access to employee benefits packages, but it would also reflect his/her years of research experience and demonstrate that the postdoc training period was effective—that this individual has established himself/herself as an independent researcher. Furthermore, by advancing that scholar, a postdoc position could now be open to a newly minted PhD who requires further mentored training as a postdoc.

By setting a term limit, institutions are essentially capping the length of postdoctoral training. How can postdocs effectively use their time to get the most out of a training experience and transition to the next stage of their career within a reasonable time (less than a term limit of 5 years)? It requires a combination of research productivity under the direction of a mentor and opportunities for career and professional development. Many institutions utilize IDPs to encourage postdocs to strategize their paths toward reaching their career and research goals. Within the IDP framework, postdocs establish their career and research goals and then identify the necessary steps they must take to reach those goals. In the process of developing an IDP, a postdoc assesses his/her current skills, abilities, and values, and then considers areas of potential growth that he/she plans to develop during the postdoc training period. According to a FASEB survey in 2012/2013, more than 80% of postdoc offices were aware of IDPs and 65% recommended the use of IDPs, yet less than 10% of postdoc offices required the use of IDPs (Hobin, Clifford, Dunn, Rich, & Justement, 2014). Significantly, only 20% of postdocs surveyed actually used IDPs (Hobin et al., 2014). More recently, as a result of NIH's Revised Policy in October 2014 that requires a description of the use of IDPs for funded trainees in NIH progress reports for T, F, and K awards (among others), more postdocs and institutions are implementing IDPs to structure their training period (Revised Policy, 2014).

By establishing career and research goals, PhDs can also leverage their postdoc position to achieve these aspirations. For example, in an academic institution, postdocs can gain valuable teaching experience toward becoming independent future faculty members. Those PhDs who know they would prefer to work in industry may want to pursue a postdoctoral position (or internship opportunity) in a company and gain valuable knowledge and experience working in that sector. Similarly, PhDs can also pursue a postdoc position in a federal research lab or apply for a highly coveted American Association for the

Advancement of Science fellowship to gain science policy experience, to name one example.

The IDP is an effective mentoring tool that can help facilitate communication between the postdoc and the mentor (Clifford, 2002). If the mentor is informed of the postdoc's intended career aspirations, together they can structure the postdoctoral training period and outline opportunities for the postdoc to develop specific skills (e.g., submitting a fellowship proposal and attending a grant writing workshop to develop grant writing skills). At the same time, the mentor can express his/her expectations of the postdoc during the training period (e.g., defining research objectives, setting work hours, and establishing regularly scheduled meetings), which can help avoid future miscommunications and/or conflicts between the mentor and the postdoc. The IDP is not intended to evaluate postdoc performance; rather, it serves as a mechanism for the postdoc to reflect on and evaluate his or her progress on an annual basis. The iterative process of self-assessment, developing a plan, and evaluating progress on a regular basis allows for the IDP to become an evolving document that will help the postdoc map out a plan for career success. Results have shown that postdocs who develop a training plan with their mentors at the start of their appointments are more likely to be successful (measured by publishing more papers and expressing greater satisfaction with their postdoctoral experience and quality of mentoring) than postdocs who do not (Davis, 2009).

As part of the IDP process, postdocs identify areas in which they intend to seek out additional training. The next section explores how institutions and postdoctoral offices provide training opportunities for postdocs to complement their research training by offering career and professional development workshops, events, and programs.

CAREER AND PROFESSIONAL DEVELOPMENT PROGRAMMING

Results from the 2005 Sigma Xi Postdoc Survey suggested that postdocs who had structured, formal training also expressed greater overall satisfaction, gave their advisors higher ratings, experienced fewer conflicts, and were more productive in terms of number of publications than those postdocs who had unstructured training (Davis, 2005). In an effort to encourage postdocs to assess their own abilities and identify areas for further development, the NPA established six core competencies (listed in Table 3.2)—or focus areas—in which postdocs are expected to develop the necessary skills and abilities to achieve professional independence and success in whichever career path they choose. The NPA defines a "competency" as an "acquired personal skill that is demonstrated in [one's] ability to provide a consistently adequate or high level of performance in a specific job function." (Rationale for Core Competencies)

Since postdocs receive much of their "discipline-specific conceptual knowledge" and "research skill development" within their laboratories and also throughout their graduate careers, postdoctoral offices focus their career and

TABLE 3.2 The National Postdoctoral Association (NPA) Established Six Core Competencies (or Areas) that all Postdocs Should Develop During Their Training Period (Rationale for Core Competencies)

NPA Core Competency Focus Area	Specific Skills Within Focus Area	Examples of Institutional Events
1. Discipline-specific conceptual knowledge	• varies based on research project and area of expertise	• gain comprehension through research/lab work
2. Research skill development		
3. Communication Skills	• writing • speaking • teaching/mentoring • interpersonal communication skills • negotiation • conflict resolution • networking • social media	• organize Annual Postdoctoral Research Symposium (National Postdoctoral Associations) • utilize American Society for Biochemistry and Molecular Biology (ASBMB) Art of Science Communication (online or blended) course (ASBMB)
4. Professionalism	• within the workplace or institution • with colleagues • connecting with society	• invite speakers, like Mary Mitchell, to lead workshops (The Mitchell Organization)
5. Leadership and Management Skills	• personnel management • project management • time management • budget management • leading a team • motivating/inspiring others	• utilize Burroughs Wellcome Fund and Howard Hughes Medical Institute's *Making the Right Moves* (Burroughs Wellcome Fund and Howard Hughes Medical Institute, 2006)
6. Responsible Conduct of Research	• nine topics established by the Office of Research Integrity (ORI) for all researchers to receive training so that research is conducted responsibly, ethically, and with the highest integrity	• utilize ORI's *The Lab* (U.S. Department of Health & Human Services) • utilize CITI Program's Responsible Conduct of Research (online) modules (CITI Program)
Additional career resources		• provided through various websites such as AAAS/Science Careers (AAAS; naturejobs.com; NatureJobs Career Toolkit; NPA, Career Planning Resources; NIH's Office of Intramural Training and Education)

Here we list the specific skills within each of those focus areas and provide some examples of workshops and professional development opportunities that institutions offer to their postdoc populations. *NIH*, National Institutes of Health; *NPA*, National Postdoctoral Association.

professional development events and programs on the other four areas by offering a variety of workshops, symposia, certificate programs, courses, and panel events to deliver this information and provide an environment in which postdocs can develop skills within these areas (leadership, management, communication, writing, negotiation, networking, and many more). Some examples of institutional initiatives, in a variety of structured learning environments (in person, online, and blended to name a few), are provided in Table 3.2. Postdoctoral associations, which are the collective body of postdocs at individual institutions, also host their own events (especially to promote a sense of community) or collaborate with postdoctoral offices to provide a wide variety of training opportunities. At institutions where no dedicated postdoc office exists, the responsibility of providing career and professional development opportunities typically falls to postdoctoral associations and/or graduate school professional development units.

Given the array of career and professional development workshops that institutions offer, it is important for postdocs to focus their activities toward reaching their research and/or career goals. Self-assessment to ascertain one's own strengths and abilities is a necessary first step. Additionally, identification of those areas that should (or need to) be developed further will help the postdoc select which workshop offerings to take advantage of. It is beneficial for this to be included into the IDP process discussed earlier in this chapter. Additionally, many institutions use self-assessment tools such as DiSC, MBTI (Myers–Briggs Type Indicator), StrengthsFinder, myIDP website (myIDP Science Careers), and SkillScan to lead postdocs through activities to evaluate their skills and innate abilities (Lundsteen, 2017). Self-assessment tools can also help postdocs identify their deficiencies (especially with respect to the list of NPA Core Competencies).

While most faculty mentors can offer advice on how to prepare for faculty positions, few are qualified or equipped to help their trainees prepare for positions outside of academia (unless they have prior experience working in industry and/or government) (Meyers et al., 2016). Through career exploration programs, postdoc offices are available to facilitate the process of informing postdocs of nonacademic career paths, including science and/or medical writing, sales and marketing, regulatory affairs, science communication, science policy, teaching, science and/or program administration, project management, and many more opportunities in the academic, government, industry, and nonprofit sectors. Career exploration programs give postdocs a chance to network and learn about various PhD career paths. These programs are synergistic with the IDP process; postdocs can gain a better appreciation for the types of positions in which their skills and strengths will be most valued, identify what aspects of the job may be challenging for them, brainstorm strategies for addressing those challenges, and access resources to advance their understanding of professional pathways.

Since most postdoc offices operate on limited budgets, many of their career and professional development events occur with minimal to no cost. Unlike training and research grants, only a few funding mechanisms exist to support these initiatives, including the Burroughs Wellcome Fund Career Guidance for Training award and, more recently, the NIH National Institute of General Medical Sciences, Innovative Program to Enhance Research Training grant. In 2013 and 2014, as a result of the NIH Biomedical Workforce Working Group Report (Biomedical Research Workforce Working Group Report, 2012), the BEST (Broadening Experiences in Scientific Training) grant mechanism provided significant funding to support 17 US universities' professional development programming (internships, workshops, and other training opportunities) over a 5-year period with the goal of transforming biomedical training by creating innovative approaches with low associated costs to prepare students and postdocs for a range of career options (as listed above).

BEST programs are still in the relatively early stages of implementation, so results and outcomes have not yet been reported. In a recent publication, however, the NIH BEST institutions identified common programmatic elements between their programs (which are also familiar to most institutions) (Meyers et al., 2016). For example, postdocs can gain valuable experience through internships/externships that potentially lead to gainful employment, but they must consider their primary obligation to the PI and their research, thus needing to balance research efforts with the time spent away from the bench. It is important to gain the support and buy-in of the faculty so that they are aware of how they can facilitate the training and/or support their trainees' engagement in external activities. Here again, the IDP can play a significant role in facilitating the discussion between the postdoc (if he/she needs an internship to help reach his/her career goals) and the PI (his/her expectations of how the postdoc will manage his/her time to still meet the research goals of the project). Furthermore, as evidenced by the T32 institutional training grant funding mechanism since December 2013, NIH recognizes nonfaculty career options as equally successful as tenure-track faculty positions; the T32 announcement clearly states that "career outcomes of individuals supported by NRSA training programs include research-intensive careers in academia and industry and research-related careers in various sectors."(NRSA). Most, if not all, of the career paths listed above are research-intensive or research-related.

With respect to faculty buy-in, one common argument against career and professional development programs is that such activities take time away from research and benchwork responsibilities. A recent report, however, demonstrated that research productivity of postdocs was increased—measured by a higher number of publications and shortened duration of the postdoctoral training period—for those who were involved in professional development programs over those who were not (Rybarczyk, Lerea, Lund, Whittington, & Dykstra, 2011). The NIH BEST awardee institutions also plan on evaluating these metrics over the course of the 5-year duration of program funding (Meyers et al., 2016). While some faculty members are very supportive of career and professional development, others are not as supportive or may be unaware of how they can support their trainees engaging in external activities. Postdoc offices continue to explore ways to engage interested faculty in discussions about postdoctoral training and professional development programs, typically through advisory committees or occasionally by presenting at departmental faculty meetings.

A major challenge that faces many institutions is to provide targeted programs for specific groups of postdocs. For example, the majority of institutions have postdoctoral populations with 50% to 60% international representation, (NSF, 2016) so there is a strong need to offer English as a Second Language courses on a regular basis. As mentioned earlier in the chapter, postdocs that are visiting the United States on visas may be disqualified from applying to many postdoctoral fellowship opportunities (those which require US citizenship/Permanent Residency status) or they cannot participate in internship/externship programs outside of their place of employment. Institutions can play a critical role by hosting networking events on campus that facilitate interaction between international postdocs and industry or government professionals.

There has been a significant increase of women and underrepresented minorities in the sciences elevating the need to offer workshops to foster the advancement of both populations and diversify the workforce (NSF, 2017). For example, the University of North Carolina at Chapel Hill has a well-established Carolina Postdoctoral Program for Faculty Diversity. Now more than 30-years old, this program has supported more than 150 scholars from underrepresented groups to transition into future faculty appointments. Professional societies, such as the Association for Women in Science and Women in Science and Engineering, often partner with institutions to offer workshops and networking events facilitating the interaction of advanced women in science with early-career scientists (Association for Women in Science).

Traditionally, postdocs in science, technology, engineering, and mathematics (STEM) fields comprised the majority of the postdoc population at any given institution; however, there has been an increase in the number of postdocs in the humanities and social sciences, and they have their own training needs and skills to develop. While many fellowship opportunities exist for postdocs in the humanities (Postdoctoral Fellowship in the Humanities), the data are insufficient about which careers humanities and social science postdocs go into and which skills are in high demand for those jobs (Segran, 2014). Institutions and organizations have recently started developing separate programs and training opportunities for postdocs outside the STEM fields (Lilli Research Group; Segran, 2014; The Versatile). In Fall 2017, the Graduate Career Consortium (GCC) will launch a new website, ImaginePhD, a career planning and exploration tool for PhDs in the humanities and social sciences (ImaginePhD). These efforts are the foundation toward developing a similar, robust catalog of training opportunities for non-STEM postdocs as currently exists for STEM disciplines.

ALUMNI DATABASES: TRACKING POSTDOC CAREER OUTCOMES

The success of institutional program efforts is highly dependent on how well they are supported by administration and faculty and whether they, indeed, add value to the postdoctoral training experience. The next section addresses tracking career outcomes as one way to start to measure the value of career and professional development programs in preparing postdoctoral scholars for future career success.

In the 1990s, 23% to 27% of science, engineering, and health postdocs attained tenure-track faculty positions (NSF, 2016). More recently, only 15% of postdocs successfully make the transition, influenced by the growing number of PhDs and the extended length of time that faculty stay in their positions, which limits the number of available jobs (Alberts, Kirschner, Tilghman, & Varmus, 2015; NSF, 2016). The fundamental question becomes, "Where do postdocs go after they leave their institutions?"

Postdoc offices are in the best position to collect data on where postdocs go after they complete their training. As a centralized administrative unit that works with human resources, individual departments, and faculty, postdoc offices track dates of employment at the institution and use voluntary exit surveys to query a postdoc's next step (Ferguson et al., 2014). Next steps include another postdoc position, a change in research status within the same institution (e.g., to staff scientist or equivalent), acquiring a new job, or, in some cases, unemployment.

A recent study from the University of California at San Francisco collected data on career outcomes for more than 1400 of their postdocs employed between 2000 and 2013. While only 37% of the group had obtained research and/or teaching faculty positions, the data showed that 81% of former postdocs were engaged in research in some capacity when nonfaculty academic, industry, and government outcomes were also included (Silva, Des Jarlais, Lindstaedt, Rotman, & Watkins, 2016). This study shed some light on career outcomes of their postdoc population; however, it was limited to postdocs supported through the NRSA T32 funding mechanism and did not include any postdocs on individual fellowships or supported through a funding agency other than NIH. The authors noted the effectiveness of tracking at the institutional level since the training environment and extent of career and professional development opportunities available to the postdocs can vary among institutions.

More recently, Melanie Sinche, Director of Education at The Jackson Laboratory, conducted a study in 2015 ("Identifying Career Pathways for PhDs in Science") to capture career outcomes for PhDs between 2004 and 2014. According to her study, only 22% of more than 3300 PhDs surveyed are currently in tenure-track faculty positions (Sinche, 2017). The other 78% are employed in various sectors (including academia, industry, government, and nonprofit sectors among others) and they are positioned in both research and nonresearch roles (Sinche, 2017). Sinche's data indicate that the majority of PhDs will successfully obtain nontenure-track faculty positions; therefore, the results support the need for graduate students and postdoctoral scholars to obtain the necessary skills and experience that can be utilized in more than just a faculty position.

Efforts to track career outcomes can lead to a better understanding of the postdoctoral training period and inform the career and professional development programming offered at the individual institutions. As noted by Kahn and Ginther (2017), nearly 80% of PhDs who invest in postdoctoral training are sacrificing sizable financial earnings and not achieving the desired outcome or reaping the benefits of earning a tenure-track faculty position. In addition to continuing to collect this valuable career outcome information, it will be important to determine how satisfied individuals are in their positions and if they feel that the time they spent pursuing a PhD is appropriately reflected in their career success. Furthermore, postdoc offices and administrators are interested in assessing the individual's postdoctoral experience, including what programs/workshops/professional development opportunities contributed to the transition from his/her training position to a job. This information can inform incoming graduate students of the career outcomes of their institution and the realities of the current climate of employment for PhDs (NIH Biomedical Research Workforce).

While there is limited available data at this time, it is clear that institutions cannot limit training PhDs to preparing only for faculty positions. That approach falls short of describing the wealth of career fields that recruit individuals with the analytical and critical thinking skills that PhDs possess. Self-assessment workshops can encourage postdoctoral scholars and graduate students to consider their strengths, skills, and values to help them identify career paths that effectively align with their strengths, skills, and values. Similarly, career exploration programs can connect postdocs and students to alumni and other PhD professionals who can share how they transitioned from a graduate program or postdoctoral position into a nonacademic faculty position.

FINAL THOUGHTS

With respect to salary and benefits, longitudinal studies will clarify if trends among institutions are consistent over time. The NPA continues to survey their institutional members to obtain information about the salary, benefits, and programs offered to their postdoc constituents. Their results will be invaluable as institutions assess if they are meeting the needs of their populations.

One aspect of the postdoctoral training experience that has been touched on throughout this chapter is the role of the faculty mentor or PI. Some PIs are supportive of career and professional development opportunities and they recognize the value of these programs (as discussed in other chapters of this book), yet these PIs may be unfamiliar with how to mentor their postdocs who want to pursue careers other than the tenure-track faculty route. Postdoc offices can serve as a resource for faculty members by providing career and professional development opportunities for their postdocs and graduate students to be successful in any career path they choose. Since the success and stability of the postdoc position is heavily dependent on the PI, it is necessary to facilitate that mentoring relationship. Formalizing the IDP and annual performance evaluation processes are ways in which postdocs and their faculty mentors can share common goals and possibly avoid potential conflicts.

While this chapter focused on the general postdoc population, it is important to address the needs of underrepresented postdoc populations. Some institutions offer fellowship opportunities, such as the UNC Carolina Postdoctoral Program for Faculty Diversity (mentioned earlier in this chapter), that promote the professional development of minority postdocs; yet additional efforts are needed to promote diversity and inclusion within the postdoctoral community and environment.

The data gathered from tracking career outcomes and how, if any, career and professional development opportunities help prepare postdocs for those careers will not only inform the postdoctoral training experience but can also transform graduate education. The BEST institutions and many others are focused on developing career and professional development workshops and events to prepare graduate students in hopes of preparing them for a wealth of career opportunities available in advance of completing their doctoral degrees. In fact, the BEST institutions developed a model to illustrate how scientific research and training in combination with career development, mentorship (by faculty and alumni), professional development, and experiential learning can enhance the graduate education and postdoctoral training experiences (Meyers et al., 2016). By preparing PhDs during graduate school, potentially fewer individuals will pursue postdoc positions unless it is part of their career plan or if they plan on leveraging their postdoc positions for a specific reason toward reaching their career goals.

Moving forward, I recommend that institutions include postdocs in the decision-making process to develop training opportunities and policies that support their needs (Committee to Review; Schillebeeckx et al., 2013). Postdoctoral offices can bring together executive leadership, faculty, and postdocs to identify needs and find appropriate solutions to help postdocs be successful during their training period and, ultimately, move on to successful careers. It will be important to ensure a continuing, open dialogue about postdoc career concerns, postdoc salaries, and benefits toward finding mutual beneficial solutions that enhance the postdoctoral training experience.

ACKNOWLEDGMENTS

The author thanks Diane Klotz, PhD (Sanford Burnham Prebys Medical Discovery Institute), Erin Hopper, PhD (University of North Carolina General Administration), and Shalini Low-Nam, PhD (University of California at Berkeley) for insightful review and editing of this chapter's content.

REFERENCES

American Association for the Advancement of Science (AAAS)/Science Careers. [online] Available at http://www.sciencemag.org/careers.

Alberts, B., Kirschner, M. W., Tilghman, S., & Varmus, H. (2014). Rescuing US biomedical research from its systemic flaws. *Proceedings of the National Academy of Sciences of the United States of America*, *111*(16), 5773–5777.

Alberts, B., Kirschner, M. W., Tilghman, S., & Varmus, H. (2015). Opinion: Addressing systemic problems in the biomedical research enterprise. *PNAS Opinion*, *112*(7), 1912–1913.

American Society for Biochemistry and Molecular Biology's (ASBMB) The Art of Science Communication online course. [online] Available at https://www.asbmb.org/Outreach/Training/ASC/.

Association for Women in Science. [online] Available at https://www.awis.org/.

Beckman Institute Postdoctoral Fellows Program. [online] Available at https://beckman.illinois.edu/research/fellows-and-awards/postdoctoral.

Biomedical Research Workforce Working Group Report. 2012. [online] Available at https://acd.od.nih.gov/biomedical_research_wgreport.pdf.

Burroughs Wellcome Fund, & Howard Hughes Medical Institute. (2006). *Making the right moves: A practical guide to scientific management for postdocs and new faculty* (2nd ed.). Research Triangle Park: North Carolina and Chevy Chase, Maryland.

CITI Program, Responsible Conduct of Research (RCR). [online] Available at https://about.citiprogram.org/en/series/responsible-conduct-of-research-rcr/.

Clifford, P. S. (2002). *Quality time with your mentor*. The Scientist. [online] Available at http://www.the-scientist.com/?articles.view/articleNo/14274/title/Quality-Time-with-Your-Mentor/.

Committee to Review the State of Postdoctoral Experience in Scientists and Engineers; Committee on Science, Engineering, and Public Policy; Policy and Global Affairs; National Academy of Sciences; National Academy of Engineering; Institute of Medicine, 2014. The Postdoctoral Experience Revisited. National Academies Press (US).

Davis, G. (2005). Doctors without orders. *American Scientist*, *93*(3, Special Suppl.). [online] Available at http://www.sigmaxi.org/docs/default-source/Programs-Documents/Critical-Issues-in-Science/postdoc-survey/highlights.

Davis, G. (2009). Improving the postdoctoral experience: An empirical approach. In R. Freeman, & D. Groff (Eds.), *The science and engineering careers in the United States*. Chicago: University of Chicago Press.

Ferguson, K., Huang, B., Beckman, L., & Sinche, M. (2014). *National postdoctoral association institutional policy report 2014: Supporting and developing postdoctoral scholars*. [online; requires NPA membership to access] Available at http://www.nationalpostdoc.org/?page=policy_report_databa.

Fishman Fund Fellowship at SBP. [online] Available at https://www.sbpdiscovery.org/support-us/projects/fishman-fund.

Hobin, J. A., Clifford, P. S., Dunn, B. M., Rich, S., & Justement, L. B. (2014). Putting PhDs to work: Career planning for today's scientist. *CBE Life Sciences Education*, *13*, 49–53.

ImaginePhD. [online] Available at www.imaginephd.com.

Kahn, S., & Ginther, D. K. (2017). The impact of postdoctoral training on early careers in biomedicine. *Nat. Biotech*, *35*(1), 90–94 (figure 1c).

Lauer, M. (2015). *Update on the postdoctoral benefit survey*. NIH Extramural News. [online] Available at https://nexus.od.nih.gov/all/2015/11/30/update-postdoctoral-benefit-survey/.

Lee, J., Williams, J. C., & Li, S. (2017). Parents in the pipeline: retaining postdoctoral researchers with families. *The Pregnant Scholar*. [online] Available at http://www.thepregnantscholar.org/wp-content/uploads/Parents-in-the-Pipeline-Postdoc-Report.pdf.

Letter from National Science Foundation's Office of the Deputy Director to the National Postdoctoral Association. [online] Available at https://grants.nih.gov/training/Reed_Letter.pdf.

Lilli Research Group. [online] https://lilligroup.com/ and Postdoctoral Fellowships in the Humanities. [online] http://www.spo.berkeley.edu/fund/hpostdoc.html.

Lundsteen, N. (2017). *Why career self-assessments matter. Inside higher ed.* [online] Available at https://www.insidehighered.com/advice/2017/01/23/using-self-assessment-tools-help-you-determine-best-career-yourself-essay.

Mason, J. L., Johnston, E., Berndt, S., Segal, K., Lei, M., & Wiest, J. S. (2016). Labor and skills gap analysis of the biomedical research workforce. *FASEB Journal: Official Publication of the Federation of American Societies for Experimental Biology*, 30, 1–11 (Figure 2c).

Meyers, F. J., Mathur, A., Fuhrmann, C. N., O'Brien, T. C., Wefes, I., Labosky, P. A., Duncan, D. S., August, A., Feig, A., Gould, K. L., Friedlander, M. J., Schaffer, C. B., Van Wart, A., & Chalkley, R. (2016). The origin and implementation of the broadening experiences in scientific training programs: An NIH common fund initiative. *FASEB Journal*, 30(2), 507–514.

The Mitchell Organization. [online] Available at https://themitchellorganization.com/.

myIDP Science Careers. [online] Available at: http://myidp.sciencecareers.org/.

National Postdoctoral Association (NPA)'s Career Planning Resources. [online] Available at http://www.nationalpostdoc.org/?page=CareerPlanning.

National Postdoctoral Association's (NPA) Postdoctoral Research Symposium Toolkit: Guidelines and Best Practices. [online] Available at http://www.nationalpostdoc.org/?postdoc_symposium.

National Research Council. (2000). *Enhancing the postdoctoral experience for scientists and engineers: A guide for postdoctoral scholars, advisers, institutions, funding organizations, and disciplinary societies*. Washington, DC: National Academy Press.

National Research Council. (2005). Optimizing postdoctoral training. In *Bridges to Independence: Fostering the independence of new investigators in biomedical research* (pp. 80–101). Washington D.C: National Academies Press (US).

NatureJobs Career Toolkit. [online] Available at https://www.nature.com/naturejobs/science/career_toolkit/cvs.

NIH Biomedical Research Workforce, Improving graduate student and postdoctoral training. [online] Available at https://biomedicalresearchworkforce.nih.gov/improve.htm.

NIH National Research Service Award (NRSA) Institutional Research Training Grant (Parent T32). [online] Available at https://grants.nih.gov/grants/guide/pa-files/PA-14-015.html.

NIH's Office of Intramural Training & Education (OITE)'s resources. [online] Available at https://www.training.nih.gov/for_trainees_outside_the_nih.

NSF. (2017). *2017 Report on women, minorities, and persons with disabilities in science and engineering*. [online] Available at https://www.nsf.gov/statistics/2017/nsf17310/digest/about-this-report/.

NSF. (2016). *Science and engineering indicators*. [online] Available at https://www.nsf.gov/statistics/2016/nsb20161/uploads/1/nsb20161.pdf (p. 361, Section 3).

Pew Research Center, Among 41 nations, U.S. is the outlier when it comes to paid parental leave. [online] Available at http://www.pewresearch.org/fact-tank/2016/09/26/u-s-lacks-mandated-paid-parental-leave/.

Postdoctoral Fellowship in the Humanities. [online] Available at http://www.spo.berkeley.edu/fund/hpostdoc.html.

The Pregnant Scholar, Funding Parental Leave. [online] Available at http://www.thepregnantscholar.org/for-postdocs/funding-parental-leave/.

The Pregnant Scholar, Title IX Basics. [online] Available at http://www.thepregnantscholar.org/title-ix-basics/.

The Pregnant Scholar, What Laws Protect Postdocs? [online] Available at http://www.thepregnantscholar.org/what-laws-protect-postdocs/.

Rationale for Core Competencies. National Postdoctoral Association. Available at http://www.nationalpostdoc.org/?CoreCompetencies.

Revised. (August 10, 2016). *Projected FY 2017 stipend levels for postdoctoral trainees and fellows on Ruth L. Kirschstein National Research Service Awards (NRSA)*. [online] Available at https://grants.nih.gov/grants/guide/notice-files/NOT-OD-16-134.html.

Revised Policy: Descriptions on the Use of Individual Development Plans (IDPs) for Graduate Students and Postdoctoral Researchers Required in Annual Progress Reports beginning October 1, 2014. [online] Available at https://grants.nih.gov/grants/guide/notice-files/NOT-OD-14-113.html.

Rockey, S. (2012). *Postdoctoral researchers—Facts, trends, and gaps*. NIH Extramural Nexus. [online] Available at https://nexus.od.nih.gov/all/2012/06/29/postdoctoral-researchers-facts-trends-and-gaps/.

Rybarczyk, B., Lerea, L., Lund, P. K., Whittington, D., & Dykstra, L. (2011). Postdoctoral training aligned with the academic professoriate. *Bioscience, 61*, 699–705.

Schillebeeckx, M., Maricque, B., & Lewis, C. (2013). The missing piece to changing the university culture. *Nature Biotechnology, 13*(10), 938–941 (Figure 1).

Segran, E. (2014). *What can you do with a humanities Ph.D., anyway?*. [online] Available at https://www.theatlantic.com/business/archive/2014/03/what-can-you-do-with-a-humanities-phd-anyway/359927/.

Silva, E. A., Des Jarlais, C., Lindstaedt, B., Rotman, E., & Watkins, E. S. (2016). Tracking career outcomes for postdoctoral scholars: A call to action. *PLOS Biology, 14*(5), 1–8.

Sinche, M. (2017). *What can you be … with a PhD?*. [online] Available at https://www.jax.org/news-and-insights/jax-blog/2017/february/what-can-you-be-with-a-phd?utm_content=buffer32fde&utm_medium=social&utm_source=linkedin.com&utm_campaign=buffer.

Stephan, P. (2013). How to exploit postdocs. *BioScience, 63*(4), 245–246.

U.S. Department of Health & Human Services, The Office of Research Integrity, The Lab: Avoiding Research Misconduct. [online] Available at https://ori.hhs.gov/thelab.

The University of North Carolina at Chapel Hill's Carolina Postdoctoral Porgram for Faculty Diversity. [online] Available at http://research.unc.edu/carolina-postdocs/applicants/.

The Versatile PhD. [online] https://versatilephd.com/.

Chapter 4

Postdoc Scholars in S&E Departments: Plights, Departmental Expectations, and Policies

Xuhong Su, Mary Alexander
University of South Carolina, Columbia, SC, United States

Chapter Outline

Introduction	69	Analyses	81
Potential Reform Efforts	71	Discussion	84
Departments as Focal Points	73	Conclusion	86
Research Hypothesis	74	References	87
Research Design	77		
Expected Career Hierarchy in			
STEM Departments	77		

INTRODUCTION

The postdoc enterprise has undergone a rapid expansion over the past few decades, the trend being particularly noticeable in science and engineering (hereafter S&E) fields (National Science Foundation, 2015; Powell, 2015). By 2015, the population of S&E postdocs was estimated at over 45,000, more than a triple increase relative to the early 1980s (National Science Board, 2016; National Science Foundation, 2015). In traditionally high-postdoc fields, such as life sciences and physical sciences, postdoc training has been a de factor career event (Davis, 2009, pp. 99–127; Su, 2013) wherein roughly 60% doctorates head for postdoctoral appointments (National Science Board, 2013; Stephan & Ma, 2005) and more than 85% of academic departments prefer new hires with postdoc experience (Association of American Universities, 2005b). Even in fields where postdoc training is not prevalent, the chances of doing such training have

The Postdoc Landscape. http://dx.doi.org/10.1016/B978-0-12-813169-5.00004-5

increased greatly (Jones, 2013). For instance, approximately 38% of engineering doctorates now head for postdoc appointments before landing permanent positions (National Science Board, 2016), with much variation also seen in other fields.

The proliferation of postdoc appointments has occurred largely spontaneously, influenced by availability of research funding and reflecting collective outcomes of discrete decisions made in academic research units or by principal investigators (Cantwell & Taylor, 2015). Postdocs are neither students nor staff on campus, whose training is dedicated to two purposes: advance scientific training and promote research careers (Association of American Universities, 1998a, 1998b; National Research Council, 1969, 1998, 2005b). Postdocs perform more than 80% of R&D in universities and are deemed critical to research excellence as "without them, research in universities would lose much of its vitality and certainly move at a slower pace" (National Research Council, 1981, p. 30). Nevertheless, with the large increase on postdoc appointments comes a great deal of concerns as documented by numerous reports (National Academy of Sciences, 2014; National Research Council, 1998, 2005a, 2005b). Though not explicitly stated, it is well expected that postdoc scholars, by investing a few years on advanced training, are to be paid off with a better career, ideally a tenure-track faculty position in Research University (Garrison, Gerbi, & Kincade, 2003; National Academy of Sciences, 2014). Yet, the fierce competition and diminishing availability of faculty positions presents almost insurmountable challenges for today's postdocs (National Science Foundation, 2015).

Approximately 80% of postdocs work at research universities, among which roughly two-thirds of postdocs aspire to seek a tenure-track position (Association of American Universities, 1998a, 1998b, 2005b). Nevertheless, most aspirants eventually end up working in other positions and sectors (National Science Board, 2016). The average duration of postdoc appointments has been increasing, as is evidenced in life sciences wherein postdoc appointments rose from 24 months for pre-1972 doctorates to the peak of 46 months for the 1992–1996 cohorts and in physical sciences rising from 21 to 30 months during the same time interval (National Science Board, 2010, 2013; Stephan & Ma, 2005). A sizeable portion of postdocs turn their appointments into "holding patterns" (Nerad & Cerny, 2002, p. VXXXV) and stay for more than 5 years (Coggeshall, Norvell, Bogorad, & Bock, 1978). Given that postdocs are hired to work on others' projects, they "spend long periods of time at the beginning of their academic careers unable to set their own research directions or establish their independence" (National Research Council, 2005a, p. 1). The fierce competition in tight academic markets further triggers concerns that postdocs may choose to work on safe bets and away from high-risk, high-reward projects to maximize their chances of landing permanent positions (National Research Council, 2005a).

These changes have significant implications not only for postdocs but also for the science enterprise. Postdocs gain less benefit from long appointments as postdoc duration is subject to the rule of diminishing returns (McFadyen & Cannella, 2004; Su, 2013). The delayed independence and lack of career prospects lead to a shred fear that promising prospective scientists may opt out for career paths that provide more chances of independence or better career outcomes (National Research Council, 2005a). This fear receives some empirical support as one survey indicates that in the biomedical field, a sizeable drop occurred in the percentage of PhDs taking postdoc training immediately after the completion of their doctoral degrees (Garrison & Gerbi, 1998; Garrison et al., 2003). Intertwined with the opting out effect is the concern for science quality. If postdoc training can no longer promise appointees good career prospects, the postdoctoral positions would be filled up with less promising candidates, who take these appointments as the best they can seek due to lack of other employment opportunities (Zumeta, 1984). Using the rankings of academic departments as proxies for quality training, multiple fields have seen substantial increase in the number and proportion of postdocs from those "unrated" departments, signaling potential concerns about the quality of scientific workforce (Nerad & Cerny, 2002, p. VXXXV). In 2008, roughly 10 percent of students indicated that they took postdoc appointments because no other appointments were available (National Science Board, 2010). The concerns for science quality may be further reinforced by the risk-averse culture prevailing in the postdoc enterprise (National Research Council, 2005a).

POTENTIAL REFORM EFFORTS

The plights in the postdoc enterprise have triggered waves of calls for policy remedies. Numerous reports (Association of American Universities, 1998a, 1998b, 2005; Committee on Science, Engineering, and Public Policy, 2010, 2014; National Research Council, 1998, 2005a, 2005b) have recommended that pertinent academic units adopt **career advising** and planning services for postdocs and enhance their opportunities for independence and different career paths. Focused on life scientists, National Research Council (2005a) urged that:

> Postdoc researchers should receive improved career advising, mentoring and skill training. Universities, academic departments, and research institutions should broaden educational and training opportunities for postdoctoral researchers to include, for example, training in laboratory and project management, grant writing and mentoring. NIH should take steps to foster these changes, including by making funds available to facilitate these endeavors.

(2005a, p. 8)

Similar statements prevail in other documents. Committee of Science, Engineering, and Public Policy (2010, 2014) proposes that postdocs should have guidance in career planning and have access to career planning offices be such agency available. Empirical studies lend support for the utilities of career advising and planning. Postdocs with a well-written career plan tend to achieve a greater level of success than those without (Davis, 2009, pp. 99–127).

With regard to prolonged postdoc training, institutions are strongly urged to adopt the **term limit** on postdoc appointments. Empirical studies find curvilinear relationships between postdoc training and knowledge creation, indicating that their productivity increases in the early stage of their appointments and decreases once appointments are long sustained (McFadyen & Cannella, 2004; Su, 2011, 2013). With multiple appointments, the cost of developing the relationships eventually outweighs the benefits of collaboration. With long sustained appointments, extended time with same collaborators produce negative impact as they eventually develop homogenous knowledge stocks and become subject to group norms and expectations (McFadyen & Cannella, 2004). The empirical turning point stays around approximately 3 years (Su, 2013). One report recommends that:

> *NIH should enforce a 5-year limit on the use of any funding mechanisms-including research grants- to support postdoctoral researchers. The nature of the position, including responsibilities and benefits, should change for those researchers who transition to staff scientist positions after 5 years.*
>
> National Research Council (2005a, p. 4)

A 5-year term limit is intended to be the maxim for postdoc positions, and it is well recommended that the normal length of postdoc training should be around 3 years, regardless of one or multiple appointments (National Research Council, 2005a). The proposal has received much institutional support. According to the reports by Association of American Universities (AAU) (1998a, 1998b, 2005b), it is common in 1998 that AAU members either have no term limit policy on their campuses or regularly ignore the policies, nevertheless, most of AAU members (74%) adopt term limit policies 7 years later. It is now more or less institutionalized in research universities due to mandates from funding agencies and advocacy from professional associations, among others (Committee on Science, Engineering, and Public Policy, 2010; 2014).

Multiple reports also point out the needs to increase postdoc salaries and provide **benefits**. Postdocs, often not classified as university staff, lack adequate access to benefits such as health insurance, leave policies, and performance reviews, among others (Davis, 2009, pp. 99–127; National Research Council, 1998; National Science Foundation, 2015). To enhance postdoc experiences, it is urged that every effort should be made to normalize the provision of medical benefits for all postdocs regardless of their sources of funding and that advisors should be asked to prepare a written evaluation of postdocs' performance at least once a year (Committee on Science, Engineering, and Public Policy, 2010, 2014).

On top of these, the science community recommends the need to collect, analyze, and disseminate **accurate information** on career prospects and career outcomes. For instance, one committee recommends:

> *Accurate and up-to-date information on career prospects in the life sciences and career outcome information about individual training programs be made widely available to students and faculty. Every life science department receiving federal funding for research or training should be required to provide to its prospective graduate students specific information regarding all predoctoral students enrolled in the graduate program during the preceding 10 years.*
>
> Association of American Universities (1998b, p. 82)

Similar calls have been documented in other reports (National Research Council, 2005a, 2005b). Gradually, the science community is building the consensus that postdoc plights require institutional intervention and possible reforms' measures need to be well implemented.

DEPARTMENTS AS FOCAL POINTS

With so many reform proposals and recommendations in the air, it is important to examine to what extent these are implemented and sort out factors that either facilitate or impede their progress. Funding agencies, professional associations, universities, and academic departments are all expected to play important roles (Association of American Universities, 1998a, 1998b; 2005b; National Research Council, 2005a). Yet, given that most postdocs are well embedded in academic departments and prove less visible in universities and other agencies, departments are the best focal points for examining postdoc plights, expectations, and practices.

Academic departments are the basic units wherein the "process of allocation of men and resources" (Zuckerman, 1970, p. 235) is unfolded and individual career prospects are shaped. Departments design rules and policies that will directly affect the long-term career outcomes for their affiliates (Ann, 2000; Fox & Colatrella, 2006; Hopkins, 2002). For instance, by mandating annual performance reviews, academic department foster mutual understanding of individual progress, which may help to prepare postdocs for job markets (Davis, 2009, pp. 99–127). With a great deal of autonomy, academic departments are empowered to leverage their resources toward certain research directions or genres, which lead to substantial influence on postdocs' research projects (Anderson, Louis, & Earle, 1994). More subtly, academic departments may establish a culture that constrains or enables individuals to pursue certain career paths and away from others (Sandler, 1986). Though academic careers are often blocked to aspirants, particularly so in life sciences where less than 10% of doctorates can succeed a tenure-track appointment (National Science Board, 2016), aspirations and aspirants themselves "are the results of an institutional training context that glorifies the academic career at the expense of other scientific career paths" (Gaughan & Robin, 2004, p. 574). Departmental culture socializes postdocs into certain behavioral patterns (Nerad & Cerny, 1999, 2002, p. VXXXV), highlighting research publications and possibly marginalizing teaching or other activities (Müller, 2014), therefore change their possibilities in pursuit of different careers.

In examining the effects of academic departments, extant literature demonstrates that individual productivity is substantially and significantly shaped by their departments (Allison & Long, 1990; Su, 2011) and that being in prestigious departments helps to have more publications, higher citations, and better career prospects (Long, Allison, & McGinnis, 1979, 1993; Long & McGinnis, 1981; Su, 2013). Two dynamics may be at work. First, prestigious departments are able to select more productive scientists, including postdocs, for their affiliates. Though decisions to hire postdocs are often made solely by individual faculty members, the flux of postdoc candidates and the tight availability of these appointments make the process highly selective (Bedeian & Feild, 1980; Burris, 2004). Second, being in prestigious departments helps individuals to have access to stimulating training, inspiring collaborators, sufficient resources, and other various incentives, all of which can be effectively factored into individual productivity (Allison & Long, 1990; Long, 1978; Su, 2011). Studies prove that postdocs from prestigious departments are much likely to succeed in their quest for prestigious appointments and prove more successful in academia over the long run (Long et al., 1979; McGinnis, Allison, & Long, 1982).

Recognizing the importance of academic departments on postdoc scholars, it is argued that postdoc plights are at least partially responsible for by academic departments wherein their expectations, policies, and rules can leverage substantial swings. If reform proposals and recommendations are to be well implemented (Association of American, 1998a, 1998b; Association of American Universities, 2005b; National Research Council, 2005a), academic departments are among the most important organizations that can either make them or break them. Studies have shown numerous cases that academic departments have a great deal of leeway to choose to enforce certain rules, safely ignore others, and manipulate different ones (Ingram & Simons, 1995; Pfeffer & Salancik, 2003; Roberts & King, 1991). Given that academic departments are often characterized by shared governance between faculty and administrators (American Association of University Professors, 1940, 1966), understanding their collective expectations for doctorates' careers is imperative to search for answers for postdocs' plights. More importantly, academic departments are the places where reforms are to be implemented and their effectiveness to be evaluated. It is important to sort out potential factors that facilitate the progress of different reform efforts.

RESEARCH HYPOTHESIS

This chapter examines to what extent academic departments implement four major reform efforts: term limit policy, career advising, health insurance, and

performance reviews, casting special attention on factors such as postdoc structure on campus, departmental resources, research strategy, and departmental
awareness, among some others. While postdocs were often seen as an invisible
college in 1970s (National Research Council, 1969), the past few decades have
witnessed their rising roles on campus (National Research Council, 1981, 2005a,
2005b). Numerous studies have proposed that institutions establish postdoc
offices and/or facilitate postdoc associations if possible (Davis, 2005; Dawson,
2007). Having such structure on campus is deemed critical to overcome decentralized management of postdoc appointments as postdoc training seems mostly
involve dyadic relationships between postdocs and their supervisors, with little
intervention from departments and less so from universities. A postdoc office
and/or postdoc association can play an important role in "consultative strategic
policymaking" (Dill & Helm, 1988). They are endowed with missions to identify issues postdocs struggle, research potential policy remedies, and implement
effective policies. More importantly, postdoc offices and/or postdoc association can serve a good communication platform for various stakeholders to share
information, develop awareness, and advocate for practices that enhance postdoc experiences (Meyer & Rowan, 1977). Once established, the presence of
such structure signals organizational commitment and constitutes a recognized
force to keep postdoc issues on policy agenda (Walsh & Ungson, 1991). Though
not necessarily confined to specific policy or practice, it is well expected that
postdoc office and/or postdoc association promotes the implementation of various postdoc-friendly policies.

**H1: Postdoc office and/or postdoc association promote the implementation
of different postdoc policies.**

Postdocs are embedded in academic departments where they conduct research,
interact with collaborators and mentors, deliver their research findings, and seek
professional support for future careers. Department chairs are key decision-makers
and arguably the most important figure for shaping departmental dynamics in policy
arenas (Carroll, 1991; Carroll & Wolverton, 2004; Wolverton, Gmelch, Wolverton,
& Sarros, 1999). Chairs exercise substantial and subtle influences on different
aspects of career incentives and are in critical positions to allocate resources. In
academia, resources are not only more than just funding but also include time,
attention, and commitment, among others. Chairs in resourceful departments
are afforded with great luxury to explore new policies and practices for fulfilling
departmental missions (Su & Bozeman, 2016; Su, Johnson, & Bozeman, 2015).
Also, sufficient resources present a strong trigger for chairs to go on uncharted ter-
ritories and pick up policies that have received little attention before (Ann, 2000).
Between different policies, departmental resources and autonomy prove to have a
direct bearing on which one is to be implemented or ignored. For instance, though
not money intensive, the resistance to do performance review is possibly rooted
in concerns for time investment, which is also a scarce resource for departments.
Extant literature provides strong evidence that in the case of diversity policies and
family friendly policies (Su et al., 2015), chairs in resourceful departments prove

to be active advocates for implementing them (Sturm, 2006). The same reasoning may likely apply to postdoc policies.

H2: Chairs in resourceful departments are more likely to implement postdoc policies.

Postdocs are critical for research excellence in academic departments. They are main producers of high-quality articles, patents, and other intellectual products (Committee of Science, Engineering, and Public Policy, 2010, 2014). Given their importance in research fields, it is expected that they are highly valued, particularly in departments committed to research strategy. After all, postdoc contribution to research progress can hardly be underestimated nor denied. It stands to reason that departments committed to research strategy may have strong incentives to implement postdoc policies. Alternatively, it is likely that academic departments perceive individual postdocs as their products (Burris, 2004). Postdocs' high research productivity are results of departmental resources, including sufficient funding, state-of-art training, and prestigious mentors, among others (Allison & Long, 1990; Long, 1978; Long et al., 1979; Long & McGinnis, 1981; Ponomariov & Boardman, 2010; Su, 2011, 2013, 2014). Good career prospects for postdocs are reflective of departmental positions in academic exchange networks (Burris, 2004). As such, departments may have little incentive to invest on postdoc policies, particularly since their appointments are short term. Those committed to research strategy may focus more on how to increase productivity with their available resources and less on policies that deviate from research endeavors. In a tight job market, the flux of postdoc candidates further helps departments to sustain their status quo as they are not incentivized to implement policies to attract more candidates, tentatively.

H3: Departments committed to research strategy are more likely to implement postdoc policies.

A large number of postdocs in the fields mean that most S&E departments have their own trained doctorates heading for postdoc appointments. When their students are more likely to pursue postdoc positions, departments may raise their awareness about postdoc plights and enhance their commitment to postdoc transition. Also, with more students heading for postdoc positions, it is likely that departments also receive sufficient number of postdoc candidates from other departments, pushing them to take measures for better postdoc experiences. Given that there is little centralized effort on many campuses to better off postdoc training (Association of American Universities, 2005a; National Research Council, 2005a), those who are struggling may be more likely to voice their concerns and dissatisfaction (Davis, 2005, 2009, pp. 99–127). Academic departments are often stressed to meet the demands from different stakeholders and may face the pressure to implement postdoc policies. The thorniest issue for postdocs is often on career prospects and their push is probably more manifested on career-related postdoc policies such as term limit policy and career advising than other ones.

H4: Departments with more doctorates heading for postdoc positions are more likely to implement postdoc policies.

Research Design

The data for this study were drawn from the 2010 Survey of Academic Chairs/ Heads. The survey targeted at the population of science, technology, engineering, and mathematics (STEM) department chairs and heads working at research-extensive universities. Out of 151 such universities, 149 STEM doctoral degree-granting universities were selected, from which all STEM department chairs were included and sent survey questionnaires. Following the steps outlined in the tailored design method (Dillman, Smyth, & Christian, 2009), the survey was administered to all subjects and out of 1832 total objects, 43% provided usable responses. Further statistical analyses show little response biases on individuals' career events or demographic characteristics.

Combined with survey data was information extracted on department prestige from *A data-based assessment of research-doctorates programs in the united states* (National Research Council, 2011). This report provides the most comprehensive information on research-doctorate programs, the programs that the survey data were specifically targeted at, and particularly ranks departments based on different productivity indicators. One concern with the NRC data is the limited size of rated academic departments. As a result, a portion of surveyed departments fail to receive departmental rankings. However, other studies have demonstrated that included chairs are not significantly different from those excluded in terms of their demographic characteristics (Bozeman, Fay, & Gaughan, 2013; Su & Bozeman, 2016), mitigating the concerns for sample representativeness.

Expected Career Hierarchy in STEM Departments

To understand the flux of postdoc scholars into the fields, it is critical to unfold how departments collectively shepherd their doctorates toward different career paths. In this regard, departmental expectations for their doctorates' career prospects matter greatly as these are the "ideas and hopes" that are passed on to doctorates and exercise substantial impacts on students' career choices (Lee, 2007). The survey asked the following question: "when you consider the kinds of positions your department's doctoral students typically obtain, please indicate which placement is preferred by the majority of your departmental colleagues?" Chairs are arguably the best individuals to answer this question and presented with the following career options for ranking: postdoc position, faculty position at a research university, faculty position in other university or college, and employed outside of academe. Fig. 4.1 shows their responses and clearly indicates that there is a preferred career hierarchy among S&E departments for their doctorates. Faculty position at a research university is the most preferred option,

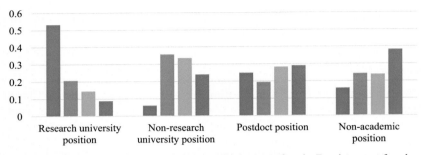

FIGURE 4.1 Preferred placements for doctorate students (percentage wise).

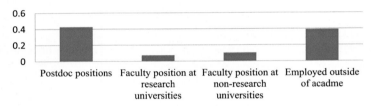

FIGURE 4.2 Positions most likely to achieve for graduating doctorates.

indicated by 53% of academic departments and the second preferred by 21%, third 14%, and fourth 9%. The pattern suggests that academic departments are more or less institutionalizing the pursuit of academic careers over other options (Gaughan & Robin, 2004). For faculty positions at nonresearch universities, few departments (6%) preferred it in the first place, but 36% and 34% prefer this option in the second and third places, possibly suggesting potential compromises with the grim realities in securing most preferred academic careers. Given that postdocs are often used as stepping stones for faculty position in research universities, it is not surprising that 25% of departments preferred their doctorates to pursue postdoc positions as their most preferred course of action, 20% as the second preferred, and 28% as the third preferred choices. For being employed out of academe, only 16% departments see it as the most preferred and relative to other options, 38% see it as the least preferred choices. Despite of grim realities that less and less doctorates are to succeed in pursuing academic careers, academic departments remain committed to glorify academic careers over other options.

To understand the extent to which their preferred options are realized, the survey asked chairs "of the positions you just ranked, which position is a graduating doctoral student most likely to obtain." Fig. 4.2 shows general patterns. About 43% of departments indicate that their graduating doctorates are most likely to obtain postdoc positions and 39% say their doctorates are most likely to be employed outside of academe. The most preferred positions

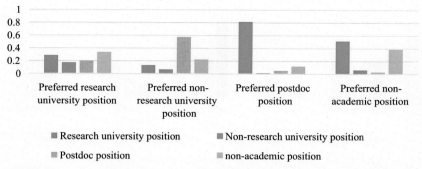

FIGURE 4.3   Preferred and achieved positions in S&E departments. *S&E, science and engineering.*

are least likely to be achieved. Only 7% of departments indicate their graduating doctorates may secure tenure-track positions in research universities and 10% for faculty in nonresearch universities. Fig. 4.3 crosstabs chairs' responses between preferred positions and achieved ones, attempting to detail differences between what is preferred and what is fulfilled. Among departments where faculty position at a research university is most preferred, 29% of departments suggest that their graduating doctorates are likely to succeed in such positions and 17% show likely success on securing faculty positions at nonresearch universities, 20% on postdoc positions, and 34% on nonacademic positions. Noticeable is that our sample is all research-doctorate departments who train the overwhelming majority of academic faculty (National Research Council, 2011). Their placements seem to deviate significantly from what is expected. For departments where postdoc positions are preferred, 81% of departments indicate their doctorates are most likely to obtain postdoc positions, and 12% to obtain nonacademic employment. Among those where faculty position at non-research universities is most preferred, 13% of departments show their students likely to secure faculty position at research universities, 6% to what is hoped for, 59% to postdoc appointments, and 22% to nonacademic positions. In just a few departments where nonacademic hiring is most preferred, their doctorates surprisingly have a good success in securing faculty positions at research universities. Given the very limited sample size for departments preferring faculty position at nonresearch universities or hiring outside of academe, it is hard to interpret their finding as either rigorous or generalizable.

Recognizing the discrepancy between what is preferred and what is achieved, the survey asked chairs to rank the following factors that may contribute to this deviation for their last doctorate who decide not to pursue the preferred placement:

- Noncompetitive salary
- Unattractive work environment

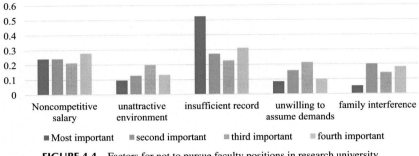

FIGURE 4.4 Factors for not to pursue faculty positions in research university.

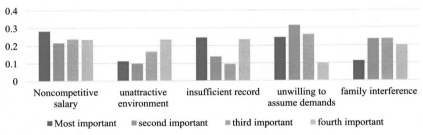

FIGURE 4.5 Factors for not to pursue postdoc positions.

- Insufficient record to warrant aspiring to such a position
- Unwilling to assume the demands of such a position
- Family interference with the demands of such a position

Fig. 4.4 presents reasons for doctorates deciding not to pursue faculty positions in research universities even though such positions are most preferred by academic departments. In the ranking order, the first and most important reason for not pursuing academic career is insufficient record, indicating the fierce competition in securing such placements, followed by noncompetitive salary, unwilling to assume demands, and family interferences. The reasons seem to indicate that pursuit of academic careers is blocked by both fierce competition in the job market and less passion among doctorates to go for the preferred courses, the latter possibly reflecting the grim realities in securing academic positions.

Fig. 4.5 presents reasons for doctorates deciding not to pursue postdocs even though postdoc positions are preferred to by their academic departments. This provides direct evidence that postdocs are struggling with different plights. In the ranking order, the reasons against pursuit of postdoc appointments are noncompetitive salary, insufficient record, unwilling to assume demands, unattractive working environment, and family interference. The findings resonate well

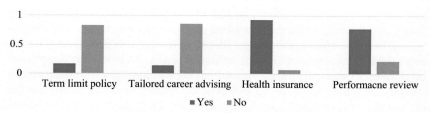

FIGURE 4.6   Department implementing postdoc policies.

with existing reports documenting postdoc predicaments (National Research Council, 2005a), suggesting that postdoc pursuit seems to be a less passionate choice among doctorates.

In summary, S&E departments maintain a well-expected career hierarchy for their doctorates, with faculty positions at research universities being most preferred choices, followed by postdoc positions, which are often perceived as stepping stones for research-oriented faculty positions. Yet, the reality challenges the realization of departmental expectations, with large number and proportion of their doctorates channeled into postdoc positions regardless of whether such positions are preferred. Nevertheless, doctorates may show less passion for pursuit of postdoc positions as they see less chances to secure academic careers and more demanding on their lives (Garrison et al., 2003; Nerad & Cerny, 2002, p. VXXXV).

ANALYSES

It is widely called that academic departments implement postdoc policies to boost their career prospects and smooth their transitions. Among all surveyed departments, 80% have postdocs in the past 3 years. Fig. 4.6 shows the percentage of S&E departments who choose to implement postdoc policies, contingent on the presence of postdoc scholars in the past 3 years. Across different policies, substantial variation is well witnessed. Only 17% of S&E departments implement term limit policies, and 14% provide tailored career advising for postdocs. Yet, noticeable is that 93% of S&E departments offer health insurance for postdocs, possibly reflecting the mandated requirements for insuring the less insured population as a part of health care reforms. 78% of academic departments provide performance reviews for their postdocs, showing high commitment toward postdocs' career development.

To further explore potential dynamics of implementing postdoc policies, variables are constructed from different sources of information. Data on postdoc offices and postdoc associations were extracted from institutional websites and often compared to the database developed by the National Postdoc Association. For department resources, the survey asked chairs whether they had full autonomy or needed approval from others on 13 items[1], including salary and funding resources, workload and working conditions, and family-related benefits. More autonomy on allocating resources is perceived as having more departmental resources at the service of chairs. An index for "departmental resources" was constructed via multiplying the dummies with their inverse weights, summing all items up and dividing further by the effective number of responses (more details, see Bozeman et al., 2013; Su et al., 2015). Chairs were asked to "please assess the extent to which you consider the following to be departmental priorities: improving the research ranking of your department, increasing the amount of sponsored research." The variable "research strategy" was constructed based on these two answers. A factor score was produced, so are the summation, both of which prove significantly related to each other. Departmental ranking can be a multidimensional concept and NRC report provides a few versions, including overall ranking, productivity ranking, and diversity ranking. The productivity ranking is constructed based on average number of faculty publications and their citations and prove to be most relevant to this study. Fields and departmental sizes are also added as control variables.

Table 4.1 shows the descriptive statistics for study variables. In 2010, roughly 40% of research universities established postdoc offices on campus and 35% showed the presence of postdoc associations. Department chairs prove to have substantial resources for running their organizations, with wide variations across institutions. The average number of publications for faculty members in S&E departments was about 1.6, consistent with existing literature (Allison & Long, 1990; Gaughan & Ponomariov, 2008; Su, 2011). Most departments perceive improving departmental ranking and increasing research funding as very important or top priorities. 33% of departments were in physical sciences, 46% in engineering fields, and the rest in life sciences. The average number of students in S&E departments was around 62.

Logistic regressions are run for different kinds of postdoc policies and presented in Table 4.2. Models suggest that the presence of postdoc offices helps academic departments to implement term limit policy, with departmental likelihood being lifted up by 40%. Those less productive departments prove more likely (10%) to implement term limit policy, arguably due to the fact that such departments need to actively attract potential postdoc candidates. After all, productivity is a strong indicator for future career prospects (Long et al., 1979; Su, 2013, 2014). Ending up in less productive departments often predicts lower likelihood of securing good academic careers in the future. Yet in less productive

---

1. The 13 items include the following: Additional salary, summer money, research money, start-up money, research assistants, course reductions, teaching assistants, computing/software, laboratory space, laboratory supplies, spousal hiring assistance, moving expenses, and travel funds. Their inverse weights are operationalized as one minus the mean of each independent variable on the ground that those less adopted merit more weights due to their scarcity and those more adopted merit less weights.

TABLE 4.1 Descriptive Statistics of Study Variables

Study Variables	Mean	Standard Deviation	Minimum	Maximum
Postdoc offices	0.41	0.49	0	1
Postdoc associations	0.35	0.48	0	1
Departmental resources	0.19	0.12	0	0.58
Research strategy	3.6	0.51	1.5	4
Average faculty publication	1.59	1.14	0.08	6.92
Department prestige	41.52	33.5	1	169
Department size	62	61	1	414
Physical sciences	0.33	0.47	0	1
Engineering	0.46	0.50	0	1
Life sciences	0.20	0.40	0	1

departments where postdocs are heavily counted on for carrying out research, postdoc policies may be perceived as effective mechanisms toward better recruitment (Committee of Science, Engineering, and Public Policy, 2010, 2014). In the meantime, in less productive departments where postdocs are more likely to have holding patterns due to lack of other employment opportunities and where postdoc plights may be very visible, departments may develop better awareness of these issues and are more incentivized to implement policies. In light of similar reasoning, those departments whose doctorates are most likely to obtain postdoc positions have a better understanding of postdoc plights and appear more likely to adopt term limit policy. More doctorate students present bigger challenges for academic departments to place them and often more obtain postdoc appointments. This likely raises departmental awareness of postdoc issues and presses departments to implement term limit policy.

For tailored career advising policies, none of prescribed factors seem to work except departmental resources. In resourceful departments, tailored career advising for postdocs are more likely to be offered, with chances being enhanced by 44% relative to less resourceful counterparts. Though multiple reports call for institutional efforts to engage in better career advising and planning (Association of American Universities, 1998a, 1998b; 2005b; National Research Council, 1998), not much has been undertaken and not many departments are serving

TABLE 4.2 Logistic Regressions on Departmental Implementation of Postdoc Policies

Variables	Term Limit Policy	Tailored Career Advising	Health Insurance	Performance Evaluation
Postdoc office	0.77(0.34)**	0.21(0.40)	0.23(0.33)	0.49(0.31)
Postdoc association	−0.28(0.34)	0.14(0.40)	0.19(0.34)	−0.47(0.30)
Departmental leaders	−0.02(0.01)	0.03(0.01)**	0.03(0.01)**	0.00(0.01)
Research strategy	−0.13(0.30)	0.19(0.37)	0.30(0.28)	0.32(0.26)
Average faculty publication	−0.29(0.16)*	−0.26(0.21)	0.12(0.15)	−0.08(0.13)
Students most likely to achieve postdoc	0.73(0.33)**	0.23(0.39)	0.95(0.37)**	−0.34(0.29)
Department size	0.01(0.00)**	0.00(0.00)	−0.00(0.00)	−0.00(0.00)
Physical science	0.41(0.42)	−0.10(0.46)	0.24(0.44)	0.50(0.37)
Engineering	−0.09(0.45)	−0.52(0.50)	0.60(0.45)	0.48(0.38)

Note: ** <0.05, * <0.10.

the calls. One critical component deterring the provision of this policy is on resources, particularly personnel and expertise to help out postdocs. Postdoc offices and postdoc associations do not seem helpful; neither does the fact that departmental doctorates are more likely to obtain postdoc positions.

For health insurance, postdoc offices and postdoc associations present no incentives or pressure for academic departments to provide such benefits. But again, in resourceful departments, health insurance is more likely to be covered than in less resourceful counterparts, the difference been up to 47% higher likelihood. Departmental size does not seem to matter, neither does departmental prestige. For performance review, it is most likely not yet institutionalized and up to discretionary decisions of mentors. None of proposed factors seem to have any significant impact.

DISCUSSION

The postdoc enterprise has been undergoing many challenges, including, but not limited to, low payment, limited career prospects, and high levels of job dissatisfaction (Association of American Universities, 1998a, 1998b; 2005a; Davis, 2009, pp. 99–127; National Research Council, 1998, 2005a, 2005b; Stephan & Ma, 2005). Over years, the science community has been wrestling with these issues and made a good many policy proposals. It is expected that academic departments where postdocs conduct research and prepare to advance their careers would take some responsibility to enhance their experiences and to improve their well-being (Committee of Science, Engineering, and Public Policy, 2010; 2014). Yet, academic departments are a part of "organized anarchy" (Cohen & March, 2000) in higher education and have a great deal of inertia and autonomy in exercising different influences on postdoc affiliates. Based on department chairs' opinion, this chapter showed illusions and disillusions about postdoc appointments, suggesting that postdoc training is largely spontaneous and lack of centralized management and that much of reform effort is driven by departmental awareness and resources rather than systematic planning. The postdoc enterprise, though much visible than before due to postdoctoral plights, remains in need to be helped in addressing various challenges.

The illusion of securing a faculty job in research universities has been sustained and engrained in academic departments (Gaughan & Robin, 2004) as about half departments indicate such positions to be most preferable. The reality, however, is that less than 10% of S&E doctorate recipients can secure such positions (National Science Board, 2016). Academic departments seem to have a good understanding of success rate for their doctorates. Only 7% of departments indicate that their doctorates are more likely to obtain faculty position in research universities and 10% of departments indicate the high likelihood to obtain faculty positions in nonresearch universities. The low success rates challenge the illusion of faculty positions being the most preferred option. Postdocs are most likely struck between glorified career options and slim chances of obtaining them, which could be a major source of stress and dissatisfaction (Davis, 2009, pp. 99–127). The illusion and disillusion are also noticeably reflected in how training is structured and managed. To pursue faculty positions, postdocs are offered little guidance about potential alternatives and are pushed often toward the only designated career option often at cost of other skills (National Research Council, 2005a). Recognizing that most postdocs would not be able to secure faculty positions, academic departments have the moral obligation to disseminate such information to their appointees and should take active measures to help postdocs toward independence, even though the paths may not be the most preferred ones.

In examining departmental commitment to postdoc policies, the most salient finding is that academic departments are not much committed to postdoc policies, particularly to term limit policy and career advising. Studies have shown that long duration is detrimental to appointees' employability and research productivity (McFadyen & Cannella, 2004; Su, 2011) and that term limit policy is urged to be implemented for both the health of postdoc enterprise and postdoc groups (National Research Council, 1998, 2005a), yet academic departments fail to grant adequate attention. In this regard, prestigious departments fail to take a good lead. On the opposite, due to their high focus on academic research and they attract more candidates than they can host, they have less incentive to adopt postdoc policies. Without isomorphic pressure (Dimaggio & Powell, 1983), academic departments are largely on their own and their implementation of postdoc policies is at discretion of their own resources and their awareness.

Departmental resources are critical to implement postdoc policies, yet this raises a thorny issue as funding for academic departments are under constant stress (National Science Board, 2016) and their autonomy is also subject to institutional-specific arrangement. It would be hard to expect nationwide progress on term limit policy or career advising. Moreover, departmental awareness seems to help to implement postdoc policies. Whether this is due to their moral pressure or serves as recruitment strategy remains to be explored. It is highly likely that both factors are at play as departments with more doctorates heading for postdoc appointments prove more committed, yet these departments are probably those who need to recruit more postdocs to perform research activities.

The recent push for organized intervention in the postdoc enterprise seems to pay off. The presence of postdoc offices on campus increases institutional commitment to term limit policy, but shows no bearing on career advising and other policies. It is likely that postdoc offices are nascent and have not yet amassed sufficient resources and support among institutions (Gewin, 2010). Given the rapid expansion on establishing postdoc offices and postdoc associations, much more institutional attention may be brought to postdoc plights and hopefully more postdoc policies can be pushed forward with these centralized efforts. The effectiveness of centralized intervention therefore warrants further studies in the future.

CONCLUSION

This chapter attempts to understand postdoc plights, departmental expectations, and policies, investigating the factors that may facilitate the implementation of postdoc policies. The findings suggest that postdoc appointees experience a great deal of illusion and disillusion and that academic departments are not much committed to implement postdoc policies. Half of academic departments see faculty positions in research universities as preferred placements, yet few S&E doctorate recipients can succeed such appointments. Postdoc appointees are struck between preferred career options and low chances of obtaining such employment, struggling with long duration, limited career advising, and other predicaments. The postdoc enterprise remains in need of help for tackling their challenges. Implementation of postdoc policies remains largely discretionary in academic departments where their resources and awareness matter and where systematic intervention falls short.

The findings have substantial policy implications for the postdoc enterprise. Academic departments have moral obligation to disseminate accurate and up-to-date information for their appointees and should be realistic about individual chances of securing faculty positions, particularly in tight job markets wherein the availability of faculty positions has significantly reduced (National Research Council, 1998, 2005a). As the discrepancy between dreams and reality gets wider, postdocs suffer more and often lead to high stress and job dissatisfaction (Davis, 2009, pp. 99–127). The bright students may opt for other career paths that provide better career outcomes and quality of scientific workforce may be compromised (Committee of Science, Engineering, and Public Policy, 2010, 2014).

For implementing postdoc policies, there is not sufficient momentum either at departments or institutions. While departmental resources and awareness help to adopt these policies, the efforts are hardly institutionalized and may often subject to ad hoc initiatives. Though postdoc office seems to push for the implementation of term limit policy, their effects are to be seen due to their nascent status. Resonating existing documents calling for more institutional commitment (Association of American Universities, 1998a, 1998b; 2005a), this study provides evidence that much of required commitment still does not exist and postdoc plights, though visible, seem not yet on the serious policy agenda. Policymakers and institutional stakeholders should be more attentive to postdoc concerns and build up problem-solving momentums such as establishing postdoc infrastructure, developing issue awareness, and seeking more resources. Future studies should provide more analyses and practical guidance along these lines.

REFERENCES

Allison, P. D., & Long, J. S. (1990). Departmental effects on scientific productivity. *American Sociological Review, 55*, 469–478.

American Association of University Professors. (1940). *Statement of principles on academic freedom and tenure.*

American Association of University Professors. (1966). *Statement on government of colleges and universities.*

Anderson, M. S., Louis, K. S., & Earle, J. (1994). Disciplinary and departmental effects on observations of faculty and graduate student misconduct. *The Journal of Higher Education, 65*(3), 331–350.

Ann, F. (2000). *Leading academic change: Essential roles for departmental chairs.* San Fransisco: Jossey-Bass Publishers.

Association of American Universities. (1998a). *Report and recommendations.* Washington, DC: Association of American Universities: Committee on Postdoctoral Education.

Association of American Universities. (1998b). *Committee on postdoc education: Report and recommendations* (Washington, D.C.).

Association of American Universities. (2005a). *Postdoctoral education survey summary of results.* Washington, D.C.: Association of American Universities: Graduate and Postdoctoral Education Committee.

Association of American Universities. (2005b). *Postdoctoral education survey: Summary of results.* Washington: D.C.

Bedeian, A. G., & Feild, H. S. (1980). Academic stratification in graduate management programs: Departmental prestige and faculty hiring patterns. *Journal of Management, 6*, 99–115. http://dx.doi.org/10.1177/014920638000600201.

Bozeman, B., Fay, D., & Gaughan, M. (2013). Power to Do…What? Department heads' decision autonomy and strategic priorities. *Research in Higher Education, 54*(3), 303–328. http://dx.doi.org/10.1007/s11162-012-9270-7.

Burris, V. (2004). The academic caste system: Prestige hierarchies in PhD exchange networks. *American Sociological Review, 69*(2), 239–264.

Cantwell, B., & Taylor, B. J. (2015). Rise of the science and engineering postdoctorate and the restructuring of academic research. *The Journal of Higher Education, 86*(5), 667–696.

Carroll, J. B. (1991). Career paths of department chairs: A national perspective. *Research in Higher Education, 32*(6), 669–688.

Carroll, J. B., & Wolverton, M. (2004). Who becomes a chair? *New Directions for Higher Education, 2004*, 3–10.

Coggeshall, P. E., Norvell, J. C., Bogorad, L., & Bock, R. M. (1978). Changing postdoctoral career patterns for biomedical scientists. *Science, 202*, 487–493.

Cohen, M. D., & March, J. G. (2000). Leadership in an organized anarchy. In M. Christopher Brown, II (Ed.), *Organization and governance in higher education*. Pearson Custom Publishing.

Committee on Science, & Engineeing, and Public Policy. (2010). *Ehancing the postdoctoral experience for scientsits and engineers: a guide for postdoctoral scholars, advisers, instutitions, funding organizations, and disciplinary societies*. Washington, D.C: National Academy Presss.

Committee on Science, & Engineeing, and Public Policy. (2014). *The postdoctoral experience revisited*. Washington, D.C: National Academy Presss.

Davis, G. (2005). Doctors without orders. *American Scientist, 93*(3), S1.

Davis, G. (2009). *Improving the postdoctoral experience: An empirical approach Science and engineering careers in the United States: An analysis of markets and employment*. University of Chicago Press.

Dawson, N. (2007). Post postdoc: Are new scientists prepared for the real world? *Bioscience, 57*(1), 16.

Dill, D. D., & Helm, K. P. (1988). Faculty participation in strategic policy making. In J. C. Smart (Ed.), *Higher education: Handbook of theory and research* (vol. IV) (pp. 319–355). New York: Agathon.

Dillman, D. A., Smyth, J. D., & Christian, L. M. (2009). *Internet, mail, and mixed-mode surveys: The tailored design method* (3rd ed.). Hoboken, N.J.: Wiley & Sons.

Dimaggio, P. J., & Powell, W. W. (1983). The iron cage revisited: Institutional isomorphism and collective rationlaity in organizational fields. *American Sociological Review, 48*, 147–160. http://dx.doi.org/10.2307/2095101.

Drotar, D., Palermo, T., & Ievers-Landis, C. E. (2003). Commentary: Recommendations for the training of pediatric psychologists: Implications for postdoctoral training. *Journal of Pediatric Psychology, 28*(2), 109–113.

Fox, M., & Colatrella, C. (2006). Participation, performance, and advancement of women in academic science and engineering: What is at issue and why. *The Journal of Technology Transfer, 31*, 377–386.

Garrison, H., & Gerbi, S. (1998). Education and employment patterns of U.S. Ph.D.'s in the biomedical sciences. *The FASEB Journal, 12*, 139–148.

Garrison, H. H., Gerbi, S. A., & Kincade, P. W. (2003). In an era of scientific opportunity, are there opportunities for biomedical scientists? *FASEB Journal: Official Publication of the Federation of American Societies for Experimental Biology, 17*, 2169–2173. http://dx.doi.org/10.1096/fj.03-0836life.

Gaughan, M., & Ponomariov, B. (2008). Faculty publication productivity, collaboration, and grants velocity: Using curricula vitae to compare center-affiliated and unaffiliated scientists. *Research Evaluation, 17*, 103–110. http://dx.doi.org/10.3152/095820208x287180.

Gaughan, M., & Robin, S. (2004). National science training policy and early scientific careers in France and the United States. *Research Policy, 33*(4), 569–581.

Gewin, V. (2010). The spread of postdoc unions. *Nature, 467*(7316), 739–741.

Hopkins, N. (2002). A study on the status of women faculty in science at MIT. In *Paper Presented at the AIP Conference Proceedings*.

Ingram, P., & Simons, T. (1995). Institutional and resource dependence determinants of responsiveness to work-family issues. *Academy of Management Journal, 38*(5), 1466–1482.

Jones, A. (2013). The explosive growth of postdocs in computer science. *Communications of the ACM, 56*(2), 37–39.

Lee, J. J. (2007). The shaping of the departmental culture: Measuring the relative influences of the institution and discipline. *Journal of Higher Education Policy and Management, 29*(1), 41–55.

Long, J. S. (1978). Productivity and academic position in the scientific career. *American Sociological Review, 43*, 889–908.

Long, J. S., Allison, P. D., & McGinnis, R. (1979). Entrance into the academic career. *American Sociological Review, 44*, 816–830.

Long, J. S., Allison, P. D., & McGinnis, R. (1993). Rank advancement in academic careers: Sex differences and the effects of productivity. *American Sociological Review, 58*, 703–722.

Long, J. S., & McGinnis, R. (1981). Organizational context and scientific productivity. *American Sociological Review, 46*, 422–442.

McGinnis, R., Allison, P. D., & Long, J. S. (1982). Postdoctoral training in Bioscience: Allocation and outcomes. *Social Forces, 60*, 701–722. http://dx.doi.org/10.1093/sf/60.3.701.

McFadyen, M. A., & Cannella, A. A., Jr. (2004). Social capital and knowledge creation: Diminishing returns of the number and strength of exchange relationships. *Academy of Management Journal, 47*, 735–746.

Meyer, J., & Rowan, B. (1977). Institutionalized organizations: Formal structure as myth and ceremony. *The American Journal of Sociology, 83*, 340–363. http://dx.doi.org/10.1086/226550.

Müller, R. (2014). Postdoctoral life scientists and supervision work in the contemporary university: A case study of changes in the cultural norms of science. *Minerva, 52*(3), 329–349.

National Academy of Sciences. (2014). *The Postdoctoral Expereince Revisited*.

National Research Council. (1969). *The invisible university: Postdoctoral education in the United States*. Washington, D.C.: National Academy of Sciences.

National Research Council. (1981). *Postdoctoral appointments and disappointments*. Washington, D.C.: National Academy Press.

National Research Council. (1998). *Trends in the early careers of life scientists*. Washingdon, DC: National Academy Press.

National Research Council. (2005a). *Bridges to independence: Fostering the independence of new investiagtors in biomedical research*. Washington, D.C: The National Academies Press.

National Research Council. (2005b). *Policy implications of international graduate students and postdoctoral scholars in the United States*. The National Academies Press.

National Research Council. (2011). In J. P. Ostriker, C. V. Kuh, & J. A. Voytuk (Eds.), *A database assessment of research-doctorate programs in the United States*. National Academy of Sciences.

National Science Board. (2010). *Science and engineering indicators*. Arlington, VA: National Science Foundation.

National Science Board. (2013). *Science and engineering indicators*. Arlington, VA: National Science Foundation.

National Science Board. (2016). *Science and engineering indicators*. Arlington, VA: National Science Foundation.

National Science Foundation. (2015). *Graduate students and postdoctorates in science and engineering*. Arlington, VA: Fall.

Nerad, M., & Cerny, J. (1999). Postdoctoral patterns, career advancement, and problems. (cover story). *Science, 285*, 1533.

Nerad, M., & Cerny, J. (2002). *Postdoctoral appointments and employment: Patterns of science and engineering doctoral recipients ten-plus years after Ph.D. completion. Communicator*, VXXXV.

Pfeffer, J., & Salancik, G. (2003). *The external control of organizations: A resource dependence perspective*. Stanford, Calif: Stanford Business Books.

Ponomariov, B., & Boardman, C. (2010). Influencing scientists' collaboration and productivity patterns through new institutions: University research centers and scientific and technical human capital. *Research Policy, 39*, 613–624. http://dx.doi.org/10.1016/j.respol.2010.02.013.

Powell, K. (2015). The future of the postdoc. *Nature, 520*(7546), 144.

Roberts, N. C., & King, P. J. (1991). Policy entrepreneurs: Their activity structure and function in the policy process. *Journal of Public Administration Research and Theory, 1*(2), 147–175.

Sandler, B. (1986). *The campus climate revisited: Chilly for women faculty, administrators, and graduate students*. Washington, DC: Association of American Colleges.

Stephan, P., & Ma, J. (2005). The increased frequency and duration of the postdoctorate career stage. In *Paper presented at the Papers and Proceedings of the One Hundred Seventeenth Annual Meeting of the American Economic Association, Philadelphia, PA*.

Sturm, S. (2006). The architecture of Inclusion: Advancing workplace equity in higher education. *Harvard Journal of Law & Gender, 29*.

Su, X. (2011). Postdoctoral training, departmental prestige and scientists' research productivity. *The Journal of Technology Transfer, 36*, 275–291. http://dx.doi.org/10.1007/s10961-009-9133-3.

Su, X. (2013). The impacts of postdoctoral training on scientists' academic employment. *The Journal of Higher Education, 84*, 239–265.

Su, X. (2014). Rank advancement in academia: What are the roles of postdoctoral training? *The Journal of Higher Education, 85*(1), 65–90.

Su, X., & Bozeman, B. (2016). Family friendly policies in STEM departments: Awareness and determinants. *Research in Higher Education, 57*(8), 990–1009.

Su, X., Johnson, J., & Bozeman, B. (2015). Gender diversity strategy in academic departments: Exploring organizational determinants. *Higher Education, 69*(5), 839–858.

Walsh, J. P., & Ungson, G. R. (1991). Organizational memory. *The Academy of Management Review, 16*, 57–91.

Wolverton, M., Gmelch, W. H., Wolverton, M. L., & Sarros, J. C. (1999). A comparison of department chair tasks in Australia and the United States. *Higher Education, 38*, 333–350.

Zuckerman, H. (1970). Stratification in American science. *Sociological Inquiry, 40*(2), 235–257. http://dx.doi.org/10.1111/j.1475-682X.1970.tb01010.x.

Zumeta, W. (1984). Anatomy of the boom in postdoctoral appointments during the 1970s – troubling implications for quality science. *Science Technology & Human Values*, 23–37.

Chapter 5

Proactive Postdoc Mentoring

Sarah C. Hokanson[1], Bennett B. Goldberg[2]

[1]*Boston University, Boston, MA, United States;*
[2]*Northwestern University, Evanston, IL, United States*

Chapter Outline

Introduction	91	Case 2: Role Models	102
What Is Mentoring?	92	Case 3: The Perpetual Postdoc	103
Characteristics and Outcomes		Case Studies: Reflection	105
of Successful Mentoring		Application of Evidence-Based	
Relationships	93	Strategies to Support Success	106
Encountering Challenges—The		Faculty Mentors	106
Complexities of Real-Life Research		Postdocs	109
Mentoring in the Academic Social		Institutions	112
System	96	Conclusion	115
Case Studies	100	Acknowledgments	115
Case 1: Discovering		References	115
Misalignment	101		

INTRODUCTION

The pillars of success within the academy—scholarship, teaching, and service—are largely social capital; they are built and maintained through productive relationships with others. Yet, on the whole, training to become a faculty member focuses more on human capital, which is attained through individual mastery of specific content within an area of scholarship rather than by demonstrating the skills needed to interact with and manage those who create it.

Mentoring postdoctoral scholars (postdocs) is a shared professional responsibility of all faculty and the institutions that support them. A good mentoring relationship is crucial to a postdoc's success in developing original research ideas and moving toward greater career independence. Successful mentoring relationships are part of a dynamic process (McGee, 2016), and the effective mentor may need to take on many roles to support the postdoc mentee—advocate, advisor, coach, and role model. Though successful interactions can take on many forms, breakdowns in faculty–postdoc relationships often happen by the same basic pathways: the faculty member and

The Postdoc Landscape. http://dx.doi.org/10.1016/B978-0-12-813169-5.00005-7

the postdoc do not trust one another or each doubts the other's commitment, expertise, and/or identity. Breakdowns in trust negatively impact the collective investment in the working relationship and can lead to lowered levels of satisfaction and productivity; in severe cases, stalled career progression for the postdoc can result.

The quality of mentoring postdocs receive has been positively associated with their ability to strategically plan for and work toward their desired career pathway (Scaffidi & Berman, 2011). Yet many postdocs report dissatisfaction with the quality of mentoring they receive from their primary faculty advisor throughout their appointment (Miller & Feldman, 2015). This is the challenge space—the gap between knowing what works and doing it. In this chapter, we explore the role of the research mentor, including the best practices identified through research on mentoring, as well as identify the cumulative advantage that a good mentor can create for their trainees. We also discuss barriers created within the academy to achieving effective mentoring relationships and examine how these challenges can create complex faculty–postdoc interactions through illustrative case studies. Finally, we provide guidelines, tools, and resources for faculty mentors, postdocs, and institutions to enable them to take a more proactive approach to building mentoring relationships that will lead to success.

WHAT IS MENTORING?

The word "mentor" originates from Homer's *Odyssey* (Homer & Fitzgerald, 1990), inspired by *Mentor* who was entrusted to look after Odysseus's household and son Telemachus while Odysseus was away. Through the responsibilities placed on Mentor—father figure, teacher, role model, adviser, protector—scholars developed mentoring as an intentional process to nurture the potential of someone junior (Anderson & Shannon, 1995; Carruthers, 1993; Little, 1990; Shea, 1997). Kram and Isabella (1985) are credited (Bozeman & Feeney, 2007) with the first contemporary definition of mentoring, emphasizing both the professional and personal support mentors provide as part of their role:

> *Mentors provide young adults with career-enhancing functions, such as sponsorship, coaching, facilitating exposure and visibility, and offering challenging work or protection, all of which help the younger person to establish a role in the organization, learn the ropes, and prepare for advancement.*
>
> Kram and Isabella (1985)

Later definitions began to shift from emphasizing the role of the mentor to defining the process itself. Bozeman and Feeney (Bozeman & Feeney, 2007) constructed their definition to be independent of organizational structures or hierarchy, focusing the inequality between the mentor and protégé on their differences in knowledge, wisdom, and experience rather than explicitly in their age or position:

> *Mentoring: a process for the informal transmission of knowledge, social capital, and psychosocial support perceived by the recipient as relevant to work, career, or professional development; mentoring entails informal communication, usually face-to-face and during a sustained period of time, between a person who is perceived to have greater relevant knowledge, wisdom, or experience (the mentor) and a person who is perceived to have less (the protégé).*

Bozeman and Feeney (2007)

Today, mentoring is neither solely defined by the mentor nor is it defined as a process focused on addressing inequalities; instead, mentoring is described as a combination of the human and social capital that leads to reciprocity and mutual benefits for both the mentor and the mentee.

> *Effective mentoring was described using the words collaborative; cooperative; confidential; confidence building; collegial; and comforting.*

Phillips (2010)

> *...mentorship ideally consists of a reciprocal, dynamic relationship between mentor (or mentoring team) and mentee that promotes the satisfaction and development of both.*

McGee (2016)

CHARACTERISTICS AND OUTCOMES OF SUCCESSFUL MENTORING RELATIONSHIPS

Mentoring relationships are generally considered effective when they result in the successful completion of career milestones and transitions and generate positive perceptions of the mentoring experience. Greater productivity and self-efficacy in a given career as well as overall career satisfaction are three of the metrics associated with positive mentoring relationship outcomes (Cho, Ramanan, & Feldman, 2011; Davis, 2005; Dolan & Johnson, 2009; McGee & Keller, 2007; Sambunjak, Straus, & Marusic, 2006). Mentees who have effective mentors have increased chances of advancement in a competitive academic landscape. Described by Merton's Matthew effect, early career advantages in academia tend to accumulate, providing future advantages to faculty as their careers continue (Merton, 1968). Postdocs are likely to seek out well-known mentors in their field and/or choose appointments at prestigious institutions because they stand to receive long-term benefits from those associations if their short-term training experiences are positive and productive (Miller & Feldman, 2015).

But there is more to mentoring than power and prestige, even if those criteria drive initial selection postdocs use to find a mentor. Mentoring models can vary in terms of their structure, function, and timing. Though the traditional definition of mentoring stems from a dyadic relationship, current models of mentoring in academia embrace networks of multiple mentors, such as peers (Kuhn & Castano, 2016; Santucci, Lingler, Schmidt, & Nolan, 2008), coaches (Williams, Thakore, & McGee, 2015), multiple advisors/principal investigators (Ensher, Thomas, & Murphy, 2001), or a hierarchal cascade (Davis, Ginorio, Hollenshead, Lazarus, & Rayman, 1996). Ideally, networked mentors would be connected and working collaboratively, though just as often they operate as independent dyads or small groups to mentor a single individual (Janasz & Sullivan, 2004). Though much has been published on mentoring relationships in academia, the diversity of training levels and disciplinary contexts makes it difficult to identify an ideal relationship model (e.g., structure, proximity, and duration) (Pfund, Byars-Winston, Branchaw, Hurtado, & Eagan, 2016).

Much of the research on mentoring focuses on the influence of effective mentoring on career choices and outcomes. A systematic review of 42 articles describing 39 studies that explored the impact of mentorship in academic medicine confirmed its value—faculty members who had effective mentors were more productive, promoted more quickly, and were more likely to stay at their institutions (Sambunjak et al., 2006). But what does "effective" really mean? Straus, Johnson, Marquez, and Feldman (2013) conducted a qualitative study across two academic health centers and found that faculty characterized successful mentoring relationships by reciprocity, mutual respect, clear expectations, personal connection, and shared values. In their Entering Mentoring training program, Pfund, Branchaw, and Handelsman (2015) defined the skills faculty and mentees should reflect and work on toward more effective mentoring: aligning expectations; maintaining effective communication; promoting mentee professional development; assessing mentee growth and understanding; maintaining equity and inclusion; fostering independence; cultivating ethical behavior; and promoting mentee self-efficacy. A randomized controlled trial conducted at 16 academic medical centers validated this workshop-based approach, and an accompanying mentoring competency assessment (Fleming et al., 2013; Pfund et al., 2013, 2014) demonstrated positive changes in faculty awareness, favorable mentees' ratings of their mentors' competency, and effective mentoring behaviors as reported by mentors and their mentees.

Our societal demographics are changing, and decades of research and funding efforts have focused on identifying strategies and interventions toward broadening participation in the scientific workforce. Recent studies have examined the role of the research mentor in the career aspirations and perceptions of PhD students (Gibbs, McGready, Bennett, & Griffin, 2014) and postdocs (Gibbs, Mcgready, & Griffin, 2015). Gibbs et al. (2015) found that the decision to pursue postdoc training was not generally associated with well-informed career intentions, and that in the absence of structured career and professional development programs, vicarious learning from research mentors shaped postdocs' career perceptions and outcomes. This unpredictable vicarious learning is particularly problematic when it comes to the career advancement of early career underrepresented minority scientists, whose interest in postdocs and academic careers declines even more than their counterparts (Gibbs et al., 2014). Successful mentoring relationships are those that are inclusive and responsive to the social and cultural identities each postdoc and faculty member bring to the collaboration (McGee, 2016; Pfund et al., 2016).

The National Research Mentoring Network (NRMN) is developing approaches to broaden participation in the scientific workforce. Recent attention has focused on new coaching models to complement mentoring, showing promise for improving underrepresented graduate students' perceptions of their ability to achieve academic (faculty) careers (Williams et al., 2015). NRMN is also developing a new training program called 6-h, to enable mentors to develop cultural awareness, designed for mentors who have completed initial mentor training and want to continue to build their skills (Byars-Winston, 2016). The intensive 6-h training program has three main objectives: facilitate participant self-reflection on their social identity to help them identify their personal assumptions, biases, and privileges that influence their mentoring style and relationships; create opportunities for the application of new knowledge through role play, group discussion, and case study activities; invite participants to develop an action or intention to become more culturally aware in their mentoring relationships, improving their ability to be more responsive to diversity issues as they may arise. This model has been pilot tested in several institutions, and NRMN is refining the preliminary evaluation data to launch a final version of this curriculum for mentors later this year (Byars-Winston, 2016).

Pfund et al. review four conceptual frameworks that have emerged in the literature that are useful in assessing the influence and effectiveness of mentoring relationships (Pfund et al., 2016)—academic persistence, social cognitive career theory, science identity, and social and cultural capital. Models of academic persistence (Chemers, Zurbriggen, Syed, Goza, & Bearman, 2011; Manson, 2009; Tinto, 1993) predict a student's completion of a degree or milestone, examining how aspects of students' social identities and lived experiences impact their ability to integrate within their institution. Social cognitive career theory (Lent, Brown, & Hackett, 1994) describes the drivers behind motivation, goal setting, and persistence toward a given academic outcome or career path, as well as how an individual's expectations and self-efficacy influence the career choices and actions they make. Science identity (Carlone & Johnson, 2007) explains how individuals develop a professional identity within the culture of their discipline, and how that professional identity is influenced both by the person's sense of recognition, performance, and competency as well as their perception of how others in the discipline view those qualities. Social and cultural capital research (Bourdieu, 1986; Smith, Beaulieu, & Seraphine, 1995) outlines the value that exists between individuals and the structures that control their access to resources and opportunity, describing the mechanisms behind how injustices are repeated over generations that allow the elite stay elite.

The synthesis of social science and anthropological theories allow researchers to develop a mentoring framework to support postdoc training programs to develop structured learning goals and activities. In their work, Pfund et al. (2016) propose a table of theory-based attributes, objectives, and assessment metrics for effective mentoring across five domains: research, interpersonal, psychosocial and career, culturally responsive/diversity, and sponsorship. Crisp, Baker, Griffin, Lunsford, and Pifer (2017) describe mentoring research and interventions for undergraduates in terms of three different frameworks: typology-related frameworks classifying how mentoring is distinct from other relationships; process-related frameworks that establish the factors that influence how mentors and mentees engage, including how their identities shape their interactions; and outcomes-based frameworks that link mentoring to student outcomes. Pfund, McGee, and others within the NRMN are in the midst of developing metrics and validated instruments to begin research studies that will, in time, enable us to more clearly articulate not just the factors that influence success, but how those factors have varying influence at each career stage across diverse disciplines (McGee, 2016; Pfund et al., 2016).

ENCOUNTERING CHALLENGES—THE COMPLEXITIES OF REAL-LIFE RESEARCH MENTORING IN THE ACADEMIC SOCIAL SYSTEM

Mentor was not chosen by Odysseus because he was proven to be an effective mentor—instead, he was chosen based on his qualifications as an elder, derived primarily from his status and privilege within the community:

> *Remember your old friend and the good turns I've done you in the past. Why, you and I were boys together.*

> Homer and Fitzgerald (1990)

In our modern higher education environment, faculty are like Mentor, entering into mentoring responsibilities based on a status they have earned through academic qualifications and their professional networks. In the *Odyssey*, Mentor must deal with the mess Odysseus left behind, and one could argue the situation prevented him from becoming as effective a mentor as the current definition of the word might suggest (Roberts, 1999)—Odysseus's palace is overrun with suitors and Odysseus's son Telemachus runs away to find his father before Mentor is able to have much influence over his growth and development. Like their namesake Mentor, faculty too have inherited a mess. Current postdoc mentoring relationships are challenged by the research landscape itself; an overabundance of postdocs and limited research funds are just two of the pressures that create tension between the need for a postdoc to produce work outputs in addition to growing and developing professionally.

Though most are well intentioned, faculty, and postdocs are fraught with many challenges within the current academic social system, creating pressure, and strain that can damage their mentoring relationships. The challenges in the postdoctoral training system have been widely documented in recent years (as identified throughout this book and (Alberts, Kirschner, Tilghman, & Varmus, 2014; Committee on Science, 2014; McDowell et al., 2014; Pickett, Corb, Matthews, Sundquist, & Berg, 2015)). Postdoc–mentor interactions have been linked to postdoc success and have also been linked to postdocs' self-reported dissatisfaction with their appointments (Davis, 2009; Miller & Feldman, 2015).

There are many cultural and structural aspects of academia, often seen through the lens of socialization (Austin & Mcdaniels, 2006), which affect the alignment between faculty and postdocs around their expectations, values, responsibilities, and practices. Research and disciplinary communities have clear expectations of what is required to contribute and belong, and part of the mentor role is to translate these implicit understandings for their mentee into explicit goals and tasks that help them advance toward long-term career success (Austin, 2002; Austin & Mcdaniels, 2006). However, this socialization that happens within departments and institutions is not generally based on evidence or research-based approaches, but on perceptions of contribution and productivity largely shaped by the mentors' experiences, and their mentor's mentor before them.

As the diversity of our graduate students increases, departments need to adapt their socialization and development processes to retain them as postdocs in the academic pipeline—meeting their needs rather than assuming they will assimilate (Williams, Berger, & Mcclendon, 2005). Byars-Winston, Gutierrez, Topp, and Carnes (2011) reported a link between perceptions of mentoring and underrepresented graduate student academic outcomes, underscoring the influence mentoring can have on underrepresented students remaining in academia. For those who do remain and enter into postdoc training experiences, we can infer the challenges underrepresented postdocs face in their academic relationships from research studying the underrepresented graduate student experience.

The primary source of both support and conflict for graduate students are their departments, graduate programs, and faculty mentors (Barnes, 2009; Golde & Dore, 2005; Lovitts, 2001), with underrepresented graduate students often having multiple experiences with subtle microaggressions, discrimination, and overt racial bias (Rowe, 1990) from peers and faculty. The inability and/or unwillingness of faculty to acknowledge race and gender during mentoring interactions of women, and women of color in particular, send direct and indirect messages that academic norms and the students' own personal values conflict (Felder, Stevenson, & Gasman, 2014; Griffin, Gibbs, Bennett, Staples, & Robinson, 2015; Haley, Jaeger, & Levin, 2014;), especially at research-intensive universities (Jaeger, Haley, Ampaw, & Levin, 2013). Underrepresented graduate students report feeling isolated (Ibarra, 2001) and frequently experience stereotype threats (Gonzales, Blanton, & Williams, 2002; Steele, 1997; Steele & Aronson, 1995), which can be compounded by lower than average expectations from faculty (Solórzano, 1993). These feelings combined with tokenization and perceived lack of respect from their academic community can lower the confidence of underrepresented students (Figueroa & Hurtado, 2013; Gonzalez, 2006) and convey messages that they do not belong (Figueroa & Hurtado, 2013; Gonzalez, 2006; Solórzano, 1993). At a minimum, to overcome these challenges within the graduate training environment, faculty need to be aware of the lived experiences of their underrepresented mentees (MacLachlan, 2006; Thomas, Willis, & Davis, 2007), the impact of microaggressions (Rowe, 1990), and the positive impact that microaffirmations can have (Cohen, Garcia, Apfel, & Master, 2006; Miyake et al., 2010).

The culture of mentoring within many disciplines is that it is a private and experiential space, a personal relationship between two people closed to the outside world. New faculty have varying degrees of experience mentoring others prior to their appointments, and varying levels of exposure to the literature and opinions on mentoring practices. The development of a faculty member's early mentoring style is informed by their experiences as a mentee, learning lessons from how their mentor mentored them as well as continuing behaviors that have enabled them to form strong relationships with other people personally or professionally. Later stage mentoring practices evolve through lessons learned through trial and error of mentoring others combined with the anecdotal experiences of their peers and colleagues. The engrained faculty scholarship practice of staying apprised of current literature in their field and synthesizing expert opinions to help inform their research directions does not extend to mentoring, much in the same way it does not extend to learning and teaching (Beach, Sorcinelli, Austin, & Rivard, 2016). Hence similar challenges exist in the field of teaching and learning, where new faculty have a range of teaching experience and do not use literature and the evidence therein as a resource when developing their own instructional practice (Beach et al., 2016).

Another challenge within the academic culture is the misalignment between postdoc career outcomes and faculty experience. Faculty are not necessarily preparing their postdocs to have the same career outcomes as them, and so experiential mentoring can fail for those not destined to become future faculty. The number of postdocs training in academia far exceeds the number of faculty positions available to them (Committee on Science, 2014), and not all postdocs desire to join the tenure track or stay in academia at all (Sauermann & Roach, 2012). A mentee's relationship with their mentor plays an integral role in shaping their career success (Austin, 2002; Austin & Mcdaniels, 2006; Jaeger et al., 2013) and forming their perceptions of their career choices (Gibbs et al., 2014, 2015). Mentors who are unwilling or lack the knowledge and/or time to effectively advise their postdoc on the availability of careers can negatively impact their progress toward achieving them. The underlying culture of academia is shifting slowly, thanks to career development programs such as the National Institute of Health-funded Broadening Experiences in Scientific

Training program (Meyers et al., 2016; Fuhrmann, 2016), but many faculty still strongly value research-intensive academic career outcomes over other career options (Gibbs et al., 2014, 2015; Sauermann & Roach, 2012), potentially limiting how effective they can be as a mentor for postdocs interested in nonacademic career pathways.

The academic system also limits mentoring relationships through the competing demands of maintaining a productive research group while fostering the relationship's growth. Expectations of faculty continue to increase, limiting the amount of time that faculty may feel they are able to set aside for mentoring postdocs. Pressure to secure and maintain research funding continues to increase as budgets tighten and pay lines get shorter, and faculty are spending more time writing, submitting, revising, and resubmitting proposals than ever before (von Hippel & von Hippel, 2015). Research administrative burdens have also increased for faculty (Gruner et al., 2015), taking their time away from other laboratory management functions such as mentoring. Though some institutions such as Cornell University (Gruner et al., 2015) have developed recommendations to streamline research administration responsibilities for faculty, shadow work, and compliance-related duties are still challenges within many research-intensive institutions.

Compounding these challenges, academia has limited rewards for being an effective mentor and limited consequences for those who are not effective. Recognition-based efforts, such as institutional awards (e.g., University of North Carolina at Chapel Hill (Anderson-Thompkins, 2016)) and external awards dedicated to excellence in postdoc mentoring (e.g., National Postdoctoral Association Mentor Award (NPA, 2017)), tend to be the primary way that faculty are rewarded for a job well done. In 2015, Purdue University approved a new promotion and tenure policy that included "an active role in mentoring, advising and supporting the academic success of students and postdoctoral scientists" as part of their review criteria (Bertoline, 2015). Though these are positive steps in affirming the importance of mentoring, they lack certain elements that could address other challenges within the training system: nominations and review are experiential based, rather than being awarded by faculty or researchers knowledgeable in the scholarship of mentoring; and the outcomes associated with them (e.g., honorarium, plaque, reimbursement of conference fees) are not motivators for faculty less invested in mentoring because they are not tied to relieving other pressures (e.g., teaching buyout, temporary extra resources, tenure, and promotion metrics).

Finally, the nature of the postdoc position itself is confusing for postdocs and faculty alike to navigate—expectations of both training and independence are entwined within the same research role, creating a paradox of autonomy (Trevelyan, 2001) that can be a hard balance for postdocs and faculty mentors to find. Further complicating this dissonance is the variability that exists among graduate student programs in the United States, let alone those globally—two entering postdocs may present similarly on paper in terms of their qualifications and their scholarly achievements yet be starkly different in terms of the level of independence they are able to maintain in their initial approach to their postdoc research project(s) based on their prior experiences in graduate school. Though lack of structured mentorship is a commonly reported mentoring challenge that limits postdoc advancement (Committee on Science, 2014; Fetzer, 2008), micromanaged postdocs also face limitations in developing the skill sets required for their next career step (Laudel & Gläser, 2008). To some extent, the culture of independence in research is just as much rooted in the faculty member's mentoring style as it is their trainees' abilities to be independent. Mismatches between those two factors during a postdoc appointment can create conflict and resentment within the mentoring relationship.

Most mentors and postdocs enter into working relationships with good intentions, but they do so within a challenging academic landscape. Social structures within training programs, experiential mentoring practices, broadening postdoc career outcomes, increased research demands, and lack of rewards can negatively affect the quality of mentoring postdocs may receive. Also, defining expectations for postdoc positions generally can be elusive for faculty, postdocs, and institutions alike, increasing the likelihood that misalignments and dissatisfaction within mentoring relationships will occur. The rest of this chapter illustrates how and why mentoring situations escalate and provides guidelines and representative resources to help academic stakeholders work together to proactively anticipate and mitigate barriers to effective mentoring relationships.

CASE STUDIES

Before we delve into evidence-based strategies that have been shown to minimize, if not mitigate, many of these challenges, we offer three case studies based on real faculty–postdoc mentoring situations to illustrate how complex and nuanced these relationships can be. We hope these case studies will be useful in helping faculty and postdocs self-reflect and have open dialogues on prior situations they could have handled differently and/or to develop skills they can apply toward actions and new approaches to mentoring relationships moving forward. In each case, postdocs and mentors both contribute to escalating the situation to a crisis point. Consider these questions as you reflect on your own mentoring relationships.

1. What are the areas of misalignment between mentors and postdocs in these cases? Have you ever experienced something similar? What strategies did you try/could they try?
2. What situations are you currently facing that might require you to reset expectations and establish clearer communication? What steps will you take moving forward?

Case 1: Discovering Misalignment

Professor Grimes is a junior engineering faculty member with an expanding research group and several new grants awarded in the last 2 years. One of his postdocs left abruptly before the end of their appointment, leaving an open position that Professor Grimes is anxious to fill. Professor Williams is a senior faculty member in the department with a long history of successful scholarship and mentoring many graduate students and postdocs. However, his research funding in recent years hasn't been stable, and his current star postdoc Mila is funded only for a few more months.

Mila was incredibly productive during her time in Professor Williams's laboratory and he is disappointed he cannot renew her contract. Upon learning that Professor Grimes has an open position, Professor Williams encourages Professor Grimes to hire Mila, offering to split her salary in the summer as a transition between the two laboratories. Mila joined Professor Grimes's laboratory, and the transition period came and went quickly. Professor Grimes and Mila continued with a normal 1-year contract, but in a few short months, the working relationship fell apart and Mila was terminated. Following her termination, Mila hired a lawyer and filed a complaint alleging wrongful termination due to discrimination.

Professor Grimes's Perspective

Professor Williams was excessive in his praise for his current postdoc Mila, describing her as hardworking and the smartest postdoc he had ever trained, able to learn anything. Professor Grimes was not sure, but Professor Williams was a prominent faculty member who had trained a lot of postdocs—his experience counted for a lot, and Grimes also did not want to rock any boat before his promotion and tenure review next year.

The initial meeting with Mila was mixed. It was hard to draw out her research expertise because she was extremely quiet and mostly looked at the floor during the interview. Halfway through the meeting, Professor Grimes noticed a picture of her children on her key chain and asked her about her family to try to bring her out of her shell, which seemed to perk her up. Professor Grimes was on the fence about offering the position to Mila, but just after her interview, Professor Williams phoned him to ask how the meeting went. Without a candidate lined up and with pressure from Professor Williams, Professor Grimes agreed to hire her.

As the fall approached, Mila demonstrated some progress in small bursts, and Professor Grimes assumed that turning to his project full-time would increase her ability to get things done efficiently. However, once Mila joined the laboratory full-time, her progress on the project plateaued even though she was now working full-time. Not only that but also she seemed to require more guidance to complete routine tasks than even some of his more junior graduate students; she was always stopping by his office. Finally, Professor Grimes terminated Mila so that he could search for a replacement that could work more independently and at a faster pace. He was surprised Mila alleged that he was discriminatory in his decision, focusing on her family situation; he felt that had nothing to do with his decision, he just needed a lot more work done than she had been able to do.

Mila's Perspective

During her initial meeting with Professor Grimes, Mila felt nervous. She relocated her family to join Professor Williams's laboratory, and she wanted this new position to work out so that her family would not need to relocate again. Midway through the meeting, Mila began to transition from feeling nervous to uncomfortable. Professor Grimes asked Mila several questions about her children, and she wondered why he was so interested and if that would bias him against hiring her. She was relieved when Professor Grimes extended her an offer.

The transition to her new laboratory was difficult for Mila. Working in both laboratories over the summer highlighted their differences. Professor Williams had made time for Mila each day she was in his laboratory, reviewing data and brainstorming about new simulations. Professor Grimes seemed put off by Mila stopping by to talk to him, so she began avoiding his office. He was also much more demanding than Professor Williams, expecting Mila to complete several tasks with minimal instructions. During the initial weeks, she had to miss a few days since her son was sick and she wondered if this was the real reason that Professor Grimes was so hard on her.

When Professor Grimes terminated her appointment after only a few months, Mila was upset and went to see Professor Williams for guidance. Professor Williams felt Professor Grimes was unfair to Mila and encouraged her to take action. Mila hired a lawyer, increasingly becoming convinced that her status as a woman and a mother may have impacted his decision to fire her.

Case 2: Role Models

Jenna is a beginning her third year as a postdoc in Professor Smith's laboratory. At the start of her postdoc, Jenna was very interested in pursuing a tenure-track faculty position, but now her interests in academia were beginning to wane. With academia seeming a more remote possibility and less than a year remaining on her postdoc appointment, Jenna decided that she needed to start focusing on her next steps.

Jenna found an internship through the City Government writing policy briefs for a local Legislator. She had always been good at writing and thought it would be a good way to explore what a non-academic career in policy might look like. The internship would require her to work in City Hall 15 hours per week, mornings from 9 a.m.–12 p.m. Monday through Friday. When Jenna approached Professor Smith about taking the internship, Professor Smith was dismissive. Professor Smith also reminded Jenna that her position was funded from her research grant, and that though she was welcome to explore policy careers on her own time, she could not take time away during the workday to pursue other jobs.

Professor Smith's Perspective

Jenna is a talented postdoc, one of the brightest Professor Smith has worked with. Professor Smith was excited to recruit Jenna and have the opportunity to mentor another strong female into getting a faculty position. It would mean a lot to her to be able to mentor someone who reminded her of herself when she was a postdoc. Professor Smith's enthusiasm has only grown over time as Jenna continues to be productive. They are close to submitting two papers, and Jenna is ready to submit a K award application for her own independent NIH funding. Professor Smith also regularly gives Jenna opportunities to guest lecture in her courses and attend conferences as part of her preparation to become a faculty member. From Professor Smith's perspective, things could not be going better. Jenna always seems assured and happy when they meet and is making excellent research progress.

Professor Smith felt blindsided when Jenna approached her with the internship opportunity in science policy. She wonders if Jenna is having second thoughts about becoming a faculty member because Jenna is truly unsure, if Jenna lacks confidence, or because Professor Smith has not been the right role model for her. However, Jenna seems to get defensive when Professor Smith reassures Jenna that she will be an excellent candidate for a faculty position. When Jenna reveals that she is interested in thinking about nonacademic careers, Professor Smith relents, but reminds Jenna that 15 h a week outside of the laboratory is not a practical expectation for someone funded on a research grant.

Jenna's Perspective

Jenna was intent about becoming a faculty member when she joined the laboratory 2 years ago, but now that future is harder and harder to visualize. Jenna has been successful in her research career to date but is burnt out from academia. She sees the way Professor Smith works at all hours and begins to realize that she is not capable of that kind of work–life balance. Though she does not really know much about careers outside of academia, Jenna knows that she needs to find out quickly if she is going to make that transition. She found an internship that she felt she could balance with her work schedule and put together a plan to discuss with Professor Smith.

Jenna and Professor Smith have always gotten along very well, so Jenna was taken aback when Professor Smith was so dismissive when she tried to talk to her about the internship. Also, Jenna resents the notion that the internship will interfere with her getting work done. She has always been productive and does not understand why Professor Smith is being so rigid about missing time during work hours—Jenna works evenings and weekends anyway.

Case 3: The Perpetual Postdoc

Daniel has been a postdoc in Professor McKnight's laboratory for almost 5 years, approaching the end of his institution's term limit for postdoc appointments, and very soon will have to identify a next step. His productivity in the laboratory has been average – he has generated just enough data and enough papers to warrant Professor McKnight continually renewing his appointment. Daniel intends to pursue a tenured faculty career, but he knows that he needs more on his CV in order to be competitive. He was unsuccessful at winning a fellowship in the first couple of years when opportunities existed, and doesn't yet have a track record of independent funding.

Professor McKnight is submitting a new grant, and Daniel would like Professor McKnight to add him to the proposal as a co-PI. If successful, it would provide longer term funding support for Daniel in the laboratory and help him submit an application to the department to become a research faculty member. Professor McKnight is less sure about committing to this long-term plan and does not immediately respond to Daniel's request. Daniel goes to meet with Professor McKnight to follow up in-person, and the conversation becomes difficult. Both sides feel the other should take more responsibility for Daniel's future success as a faculty member.

Professor McKnight's Perspective

Professor McKnight has not known what to do with Daniel for a long time. It has been clear to Professor McKnight for almost 2 years now that Daniel does not have what it takes to become a tenure-track faculty member, even though he continues to work toward that career. Still, Daniel has always asked for his appointment to be renewed each year, and Professor McKnight has always agreed. Daniel is productive enough that he is worth having around—Daniel frequently trains new graduate students, and he does generate results.

Daniel's term limit ending coincides with the submission of a new grant proposal that builds on a project Daniel has been working on for several years. Professor McKnight could see Daniel continuing the work, but even as co-PI, Daniel would be in limbo. Professor McKnight truly does not think Daniel would develop the independence or drive he would need to gain additional independent support and move beyond this project in Professor McKnight's laboratory. Professor McKnight also does not feel comfortable putting Daniel up for a promotion to research assistant professor because he is not sure that his fellow colleagues would be supportive. It might be better for Daniel to find a new position and a fresh start somewhere else, but Professor McKnight is not sure yet how to approach that conversation with Daniel. Professor McKnight is caught off guard when Daniel stops by his office to follow-up, and he pieces together his feedback awkwardly to finally tell Daniel that he is likely not going to be successful in pursuing a faculty career. He feels guilty when Daniel gets upset; he is not exactly sure at what point he should have provided more feedback, but he guesses there was one. So many of his other postdocs succeeded without much intervention from him—he assumed Daniel would be similar.

Daniel's Perspective

Daniel has known that he has wanted to be a faculty member since his days as an undergraduate researcher—he has never even considered the possibility of a different career path because the faculty track is such a good fit for his intellectual curiosity and laid-back working style. Time has flown by faster than Daniel had even kept track of, and now 5 years into his postdoc, he still is not where he needs to be relative to his friends who have transitioned to faculty positions successfully. At this point, Daniel thinks his best chance for success is to stay within the department, transition into a research faculty appointment, and work to get his own funding that will launch his independence. Professor McKnight's upcoming proposal is an opportunity to make that happen—Daniel can continue working on his project with longer term funding support, and he will obtain the PI-ship he needs to get promoted and be more competitive for other grant opportunities. Daniel does not see a downside—Professor McKnight has been happy enough with his work to keep him around this long.

Daniel is put off when Professor McKnight avoids his email request and so goes to stop by to confirm he can help lead the proposal. He is even more frustrated and caught off guard when Professor McKnight instead suggests that Daniel begin to think about finding another position outside of the laboratory. Daniel has invested a lot of time in Professor McKnight's laboratory, and he feels he has not always gotten the same investment back from Professor McKnight. Professor McKnight has not gone out of his way to help Daniel become more successful. In particular, Daniel thinks that his fellowship applications would have been stronger if Professor McKnight had spent more effort providing feedback. Daniel leaves the conversation without a clear resolution of what is next. He is sure he could find a second postdoc, but unsure that move would boost his already slim chances of obtaining a faculty career.

Case Studies: Reflection

Each of these cases highlights a particular situation that has magnified how a mentor and their postdoc can be disconnected from one another. By imagining themselves in the identities of the characters in these cases, faculty and postdocs can reflect on their own approaches and identify where they may need to adjust to build stronger working relationships. Proactive faculty and postdocs can frame their relationship in terms of their aligned values so that they are better equipped to navigate some of the challenging situations and conversations that they may encounter. In the next section, we will highlight the evidence-based strategies faculty and postdocs can apply to support their success, as well as approaches institutions can take to foster supportive mentoring relationships.

APPLICATION OF EVIDENCE-BASED STRATEGIES TO SUPPORT SUCCESS

Effective mentoring relationships are not passive but are instead active exchanges that foster collaboration between the faculty member and the postdoc (McGee, 2016). There are steps that faculty, postdocs, and their institutions can take to ensure greater success. The solutions and resources provided here are not exhaustive but instead are meant to be representative examples of evidence-based approaches that could be proactively used to mitigate or minimize the challenges described above.

Faculty Mentors

The ability to set expectations is an important leadership skill, but many faculty members struggle to clearly explain how their research environment functions to their mentees. Often, the socialization of new postdocs into their new training environment is more implicit than explicit, and a postdoc's primary understanding of the political, ethical, economic, and social dynamics within their academic community is developed through their own lens rather than through the input from the perspective of their mentor (Miller & Feldman, 2015). This style can work when the faculty member and postdoc intrinsically share similar perspectives and personalities, but if not, it can result in loss of trust, miscommunications, and frustrations for both the faculty member and the postdoc.

Research on social identity and behaviors in groups (Tyler & Blader, 2003) has demonstrated that the more strongly people identify with a group, the more effort they put into working toward its mission and achieving mutual goals. All team members benefit from having a clear sense of what is expected within a given work environment (Lencioni, 2002); postdocs should understand what they can expect from their mentor, and faculty members should be clear about what they expect from their postdocs. Following through on these expectations builds a level of trust over time that can allow faculty and postdocs to approach more challenging and unexpected situations successfully. Mentoring compacts are one tool that mentors and postdocs can use to establish their expectations and build trust at an early stage (AAMC, 2017), trust that is then reinforced as expectations are met and refined throughout a postdoc's appointment. Giving postdocs a "Welcome to My Lab" letter is one example of a mentoring compact framework with which mentors and postdocs can develop working relationships and build trust (Bennett, Maraia, & Gadlin, 2014).

Letters generally cover the expectations for both the faculty mentor and the postdoc during the relationship and should take into consideration that postdocs come from a wide variety of cultural and training backgrounds. Faculty should not assume that postdocs understand the norms within their institution, department, or research group; letters should cover the basic work expectations (e.g., hours committed, time overlapping with faculty member during working hours, professional conduct) needed for the postdoc to be successful.

Additional topics included within the letter could include

- **General expectations:** goal of the research group, role of the faculty member and other leading members of the team, expectations of research group members
- **Expectations for team interactions:** team structure and reporting/supervisory roles, team meetings, journal clubs, sharing space and facilities, time and attendance, vacations and leave, networking and attending outside meetings, professional etiquette, expectations for collaboration with and training of other group members, expected work habits, faculty member's work habits
- **Expectations for collaboration:** description of external collaborations and expectations for the role of the postdoc in the project(s)
- **Responsible conduct of research:** research integrity, required record keeping, and data sharing practices, definition of reproducibility within the research environment, institutional guidelines, and required trainings, resources to report research misconduct
- **Communication:** preferred modes of communications, preferred style of meetings (e.g., scheduled with agendas, informal), process to follow if there is a disagreement
- **Work style:** turnaround time for emails or items to review, best times of day to reach faculty member
- **Authorship and acknowledgments of scholarly contributions:** criteria for deciding order of authorship or credit for scholarship, process for manuscript preparation and submission, other ways credit will be acknowledged for work contributed (e.g., talks, posters), guidelines for seminars/talks
- **Proposal writing:** expectations for individual fellowship applications, expectations for contributions to faculty member's proposals
- **Evaluation and feedback:** form of feedback (e.g., performance review, individual development plan (IDP)) and frequency, process for obtaining reference letters
- **Mentoring:** expectations and style of faculty member, expectations for postdoc to mentor others in the group, expectations for how/if mentoring contributions are acknowledged
- **Career and professional development:** time committed to professional development, opportunities for professional development within the institution and externally
- **Institutional and local resources:** contact information for postdoc office, departmental administrators, human resources, international scholars' office, or other support structures within the University that assist postdocs

Imagine Mila had received a "Welcome to my Lab" letter from Professor Grimes. She would have known about his preference for independent work versus many face-to-face meetings, and she may have asked Professor Williams to help her find a lab that was better suited to the style of mentoring and feedback she needed to be successful. Or, she could have asked Professor Grimes or other members his group if there were opportunities to work collaboratively, building a network of peers to help her get more familiar within the group at her start. Either way, the letter would have provided a structure for them to clarify their expectations; without the letter, neither Professor Grimes nor Mila took the opportunity and both made assumptions that damaged their relationship.

These letters and other forms of mentoring compacts are most effective when they serve as conversation starters rather than the end of these discussions. Providing the opportunity for postdocs to input their own goals and expectations into the mentoring compact creates buy-in and belonging that will increase their motivation to perform as part of the faculty member's team (Tyler & Blader, 2003). Once a postdoc has assimilated into a research group, the contents of a compact or letter can be translated into an IDP. IDPs allow the postdoc to consider their current skills, interests, and values to assist them in developing career goals alongside their faculty mentor (FASEB, 2003; Fuhrmann, Hobin, Lindstaedt, & Clifford, 2015). Career planning tools provide value because they ensure that mentor and postdoc expectations continue to align throughout the appointment, and such tools include an assessment of the postdoc's skills and progress, identification of research- and career-related goals, and action items for the postdoc and the mentor to guide future meetings. The benefits to the postdoc are clear—the 2005 Sigma Xi Postdoc survey of US postdoctoral scholars showed that postdoctoral scholars who created a written career plan or IDP with their mentors were 23% more likely to submit papers, 30% more likely to publish first-authored papers, and 25% less likely to report that their mentor did not meet initial expectations (Davis, 2005). However, IDPs have significant benefits to faculty as well, providing a framework for career discussions and performance feedback that can often be difficult to integrate into regular research progress meetings (Fuhrmann, 2016).

Daniel and Professor McKnight would have benefitted from using a career planning tool such as an IDP to help Daniel monitor his research progress and align his short-term goals with long-term career preparation. Professor McKnight was an untapped resource for Daniel—their conversations did not seem to make the most of his expertise or the network of former postdocs that had gone on to successful careers. Reviewing an IDP regularly might have helped Professor McKnight identify opportunities to give Daniel feedback at an earlier stage, providing Daniel with insight to areas where Daniel needed to be more proactive to remain competitive with his peers. Based on this feedback, Daniel might have succeeded earlier in his postdoc, or he might have been able to identify other career opportunities or another postdoc position where he could be more successful.

Given the power dynamic created by the supervisory role faculty have over the postdocs who they mentor, the responsibility for creating an open environment to share expectations and feedback on a regular basis falls to the mentor, who should lead by example. Setting clear expectations and maintaining clear lines of communication can overcome many of the challenges faculty and post-docs may face together, even if these strategies alone are not enough to avoid every challenging situations. Establishing expectations that are explicit and well understood increases the trust created when those expectations are achieved.

Postdocs

In this section, we apply the principles of being proactive, developing open communication, and building and following through on expectations to postdocs—the other party in faculty–postdoc mentoring dyads. Though their mentor will contribute to their career advancement and success, the postdoc also bears responsibility for and ownership of their own career path. In truth, the most successful mentoring relationships are those in which the mentee takes initiative and drives the mentoring partnership to fulfill their needs (Committee on Science, 2014; Fuhrmann, 2016; Miller & Feldman, 2015; Scaffidi & Berman, 2011; Su, 2011).

Postdocs often believe that their mentoring relationship starts when they accept a job offer, but it can start much sooner—during the interview. Interviewing with a research group is the postdoc's first opportunity to understand group norms and expectations and decide if those align with their own goals and expectations. Though in today's competitive academic research landscape the prioritization is often on securing a job and less on evaluating the job itself, postdocs can and should proactively choose groups and mentors who align well not only with their scientific interests and desired career outcomes but also with their own work style and professional goals.

In our experience, most of the time graduate students invest in postdoc interview preparation is generally spent reading recent publications from the research group and preparing the graduate student's research talk reflecting their thesis work. This type of preparation focuses solely on scholarship content and does not help the future postdoc assess the suitability of the research environment as a whole. Self-reflecting on their expectations as well as asking questions during an interview about the expectations and norms of the research group will identify points of alignment and misalignment, as well as allow for an assessment of the overall environment of the research group.

Though the goal should be to identify postdoc opportunities where the potential postdoc's expectations are as aligned with the mentor's expectations as much as possible, there will be points of divergence within this set of questions recommended below and any other aspects of the position the postdoc will consider. As part of this reflection exercise, postdocs should also consider which questions and issues are the highest priority for them now, and they should anticipate what they think their priorities might be 1 year into the appointment. This will ensure that postdocs find alignment in the issues that matter most to them, which will hopefully make them more willing to find compromise in areas that are less important. Table 5.1 provides sample self-reflection questions

TABLE 5.1 Finding Alignment—A Checklist for Postdoc Interviews

Questions for Self-Reflection	Questions for Faculty Mentor/ Research Group
What are my research interests in this laboratory? What skills do I want to develop?	How will my project contribute to the overall goals of your laboratory?
What are my top three priorities during my postdoc training?	What do you expect from postdocs who join your lab? What are your top three priorities for training?
What kind of job stability do I need?	Is this position renewable? How long is there funding to support this position?
What are my career goals right now? What support will I need to pursue that career?	What careers have other postdocs from your lab pursued? What resources did they have?
What professional development opportunities (e.g., conferences, workshops) do I expect?	What professional development opportunities do members of your lab routinely take advantage of?
What does productivity mean to me? What do I want to get out of my postdoc training for my CV?	What are your expectations related to productivity? What have past group members produced?
What are my expectations related to publication? What contributions do I think merit authorship?	What are your expectations related to publication? What contributions merit authorship in your laboratory?
What are my goals for proposal writing? Do I expect to write my own fellowships? Do I expect to contribute to larger grants?	Will I be able to write my own fellowship awards? What is your expectation related to collaborating on larger proposals?
How do I like to work? Am I more independent, or do I perform better with regular guidance and feedback?	What is your mentorship style? How frequently do members of the lab meet with you?
How do I like to communicate? When I am most comfortable?	What style of communication do you prefer?
How do I give and respond to feedback?	How do you give and respond to feedback? What is the best way to approach you?
Do I want to mentor other students during my training? Will I be mainly providing support/advice or contributing to their research?	What are your expectations for mentoring in the lab? How are research contributions in mentoring situations weighed in terms of authorship?

TABLE 5.1 Finding Alignment—A Checklist for Postdoc Interviews—cont'd

Questions for Self-Reflection	Questions for Faculty Mentor/Research Group
Do I want to gain additional experience (e.g., teaching) during my postdoc?	What opportunities are there for me to build skills outside of those I will learn and apply in your group?
Is a sense of community important to me?	What is the social environment like in this laboratory? How can I meet other postdocs in the department or across the institution?
Do I have any hobbies that are a priority to me?	What resources are there for me to continue doing my hobbies in my free time?

as well as sample questions graduate students can ask their potential faculty mentor and/or members of their laboratory.

This checklist does not contain questions related to personal or family issues, though these are often a consideration in choosing a postdoc appointment. Job seekers have different areas of comfort in raising these issues during the interview process, but if a personal or family consideration is a "dealbreaker," then it is important for the postdoc to discuss that situation with the faculty member before accepting a new position. Some postdocs may feel comfortable raising personal or family issues in the interview setting, while others may choose to wait until an offer has been extended to follow-up. An advantage to bringing up these issues before or during the interview is that the institution may have points of contact within other offices or resources that the postdoc could meet with during their visit.

It is unlikely that one mentor will fulfill all of a postdoc's specific needs (Janasz & Sullivan, 2004). Rather than accepting a lack of mentorship in specific areas or skills, postdocs should also identify and seek out other mentors to meet with regularly. A multimentored postdoc can capitalize on the unique skills of many individuals, cross-training across disciplines, sectors, and competencies. Postdocs can learn from the model developed by de Janasz and Sullivan encouraging multiple mentors for junior faculty members—identifying mentors based on their beliefs and identities, knowledge, and skills, and networks or external relationships (Janasz & Sullivan, 2004). This group of mentors may or may not be in the postdoc's institution, academic discipline, or even in academia at all, and they may change as the postdoc evolves throughout their appointment.

Jenna's case study highlights one pathway by which postdocs become drawn to nonacademic careers, but it does not mean to imply that one has to be burnt out from academia to determine that a career outside of the academy is right for them. It is hard to say from the information in this case why Jenna was drawn to a career outside of academia and because Jenna was not having regular conversations with her mentor about how she was feeling throughout her appointment, Jenna likely is not completely sure either. It could be that Professor Smith was not the right role model for Jenna, and a role model more aligned to Jenna's way of working might have helped her envision staying on an academic path. Or, it could be that Jenna would have found her way out of academia even with a faculty role model more like her—maybe policy is just truly the right fit, and it took a longer time working in academia for Jenna to realize that. But, if Jenna had developed a network of mentors outside of Professor Smith earlier in her postdoc, she might have been more prepared to explore her career choices earlier and might have felt more supported in her intense working environment.

The strategies we suggest here provide early opportunities for postdocs to take charge of their own mentoring relationships. However, being proactive as a postdoc means being an active partner even after the relationship is established, including taking the following actions: accepting responsibility for actions (present and future); building strong relationships with others to fulfill mutual goals; respecting a mentor's time; establishing clear lines of communication; accepting criticism and feedback constructively; and playing an active role in solving problems (Muller, 2009).

Institutions

Institutions share responsibility in supporting successful faculty–postdoc relationships by creating systemic changes that prioritize and incentivize good faculty mentoring behaviors and practices. Ultimately, while individual faculty members shape the environment of their research group and can influence the culture of their department, institutions must address the broader opportunities, abilities, and motivations (Rothschild, 1999) required to achieve larger shifts in culture over time (Henderson, Dancy, & Niewiadomska-Bugaj, 2012; Wieman, Perkins, & Gibert, 2010).

One challenge institutions can address is the lack of professional development opportunities to support the mentoring relationships between faculty mentors and their postdocs. Faculty are essential to creating culture change within higher education, but for many, the only context and model for their mentoring is experiential rather scholarly. Faculty need intentional opportunities for self-reflection combined with professional development opportunities to become proficient in mentoring skills. The current practice of assuming faculty members and postdocs will enter into productive relationships on their own, without resources and interventions provided by their research institution, favors postdocs who have already developed or innately possess the social skills to create connections with their particular faculty mentor (Pfund et al., 2016).

Though the "Welcome to My Lab" exercise described above is a useful self-reflection tool, it is still a tool representing the faculty member's own personal approaches to and perspectives on mentoring rather than reflecting evidence-based practices. Institutions can help faculty become aware of scholarly-based approaches to research mentoring through establishing regular training opportunities, drawing on the resources developed through the Entering Mentoring program at the University of Wisconsin–Madison (Pfund et al., 2015) or National Institute of Health-funded programs such as the NRMN and Clinical and Translational Science Institutes. Many institutions currently offer research mentor training support through optional "one off" faculty development opportunities through a workshop or series of workshops. To affect change, research mentor training should shift to become an expectation rather than a choice, and it should be habitual rather than a single, check-the-box activity completed and done with. Research-active faculty should have access to training materials and resources on a renewing basis to provide a framework for them to be regularly conscious of their own progress as a mentor and allow them to have access to and adopt evidence-based mentoring practices as they continue to emerge in the literature. Institutions can reinforce these resources for faculty by also offering professional development for postdocs that empowers them to take ownership of their career advancement.

In addition to the foundational mentoring skills described above, professional development opportunities for postdocs should help them identifying their needs and learn the advocacy skills they will need to gain sponsorship and support from their mentor, adopting "mentoring-up" approaches similar to those employed at the University of Wisconsin–Madison and Northwestern University (Lee, McGee, Pfund, & Branchaw, 2015). Part of institutional support for "mentoring-up" could be adopting policies that provide consistent access and adoption of career planning tools such as IDPs (FASEB, 2003; Fuhrmann et al., 2015) across all disciplines and postdoc training programs. Career planning tools are linked to postdoc career success and satisfaction (Davis, 2005), and they are a powerful way for postdocs to develop more agency in their own career and within their mentoring relationships.

Additionally, most institutions have or are now developing career development resources to support how faculty help postdocs prepare for their next career step, building off of the success of programs like those at the 17 NIH-funded BEST institutions. Gibbs et al. recently studied the career development of postdocs and the efficacy of previously described success factors as predictors of postdocs' interest in specific career pathways (Gibbs et al., 2015). They found that in the absence of structured career and professional development programs, vicarious learning from research mentors shaped postdocs' career perceptions and outcomes. It is important that what postdocs pick up vicariously from their mentors be well informed. Institutional leaders responsible for professional and career development opportunities for postdocs must recognize the importance of including faculty in at least supporting their postdocs to attend career development opportunities, and even better, to get faculty engaged themselves. Our experience has shown us that faculty are more likely to participate under the following conditions: there exists a low level of engagement, for example as an occasional cofacilitator; one of their current or former students is already participating; they see tangible benefits to their research program (e.g., better performance, career success, and recruiting); and/or they are members of large-scale postdoc training grant programs and are committed to sharing the administrative and training loads in professional development.

Institutions must also remain committed to developing and implementing ongoing strategies related to increasing diversity and inclusion. While over half of PhD students begin graduate school with the career goal of obtaining a faculty position (Golde & Dore, 2001), one-third lose their interest in faculty careers during their PhD and half lose an initial interest in becoming faculty at a research university (Fuhrmann, Halme, O'sullivan, & Lindstaedt, 2011; Mason, Goulden, & Frasch, 2009; Sauermann & Roach, 2012). For underrepresented students, this decrease in interest is 50% larger than the corresponding drop for majority males, the group whose interest remains the highest (Gibbs et al., 2014). Many underrepresented early career researchers have difficulty envisioning themselves in academia as faculty because they do not have access to mentors with similar social identities. Gibbs, Basson, Xierali, and Broniatowski (2016) demonstrated that the postdoc to faculty transition is the biggest barrier to advancing underrepresented groups within academia; despite dramatic increases in diversity within the PhD student population, researchers from underrepresented backgrounds are not being hired into faculty positions at the rate needed to establish parity within medical school basic science departments. Institutional efforts and resources directed at improving the climate within academia, such as those published by Gutierrez et al. (2014). and the culturally aware mentoring programs developed by the NRMN (Byars-Winston, 2016), can be coupled with intentional recruitment and retention programs at all stages of the academic pipeline, increasing the strength of underrepresented postdoc–faculty relationships over time.

In the longer term, strategies to incentivize and reward good mentoring practices will reinforce faculty buy-in and participation in interventions such as those outlined above. Defining institutional standards for mentorship based even on the current understanding of metrics for mentoring success would be a powerful first step. Over time, institutions recognizing successes and creating expectations for avoiding failures will motivate faculty to work to establish new social norms for their departments and research groups. Mentors who take pride in their mentoring often reference their positive feelings about supporting their trainees' successes as an internal driver for their time investment. Institutions may not need to develop external motivators as much as ensure that the professional development opportunities they develop for faculty activate and enhance the intrinsic motivations within faculty to be good mentors. Recent work by the NRMN demonstrates that when faculty develop mentoring skills coupled with awareness of the integral role mentoring can play in sustaining effective work teams, the majority of faculty will modify their behaviors toward practices that support success (McGee, 2017).

CONCLUSION

Cumulative advantage in academia begins as early as the graduate and postdoctoral training stages, where opportunities for success are influenced by whether trainees have access to engaged, positive and supportive mentoring relationships, and carry on throughout their development. Critical aspects of positive postdoc–faculty mentoring relationships include establishing expectations, facilitating clear communication, fostering independence, and creating inclusive research environments. As postdoc–faculty relationships have been demonstrated to influence the postdoc's career satisfaction and success, the future of the research workforce depends on getting these relationships (and those within the postdoc's wider mentoring networks) right.

We recognize that longer term changes within the academic social structure are hard to achieve, and that the solutions and applications we present here are not necessarily designed to address the systemic problems at large. However, we hope that the resources and case studies in this chapter are valuable for individual faculty and postdocs in the context of future professional development opportunities, as they are reflective of the types of situations they will likely encounter and potentially can overcome together. Improvements in individual postdoc mentoring outcomes will largely depend on the ability of faculty and postdocs to shift their approach to mentoring relationships from experiential to scholarly, being more reflective of their own practices and adopting evidence-based practices proactively. Institutions can help them by creating an infrastructure that encourages self-reflection by both mentors and their mentees, thus generating the awareness that leads to the adoption of the evidence-based practices needed to ensure mentoring success.

ACKNOWLEDGMENTS

The authors acknowledge Rick McGee and Gary McDowell for providing feedback during the preparation of this submission.

REFERENCES

AAMC. (2017). *Compact between postdoctoral appointees and their mentors.*

Alberts, B., Kirschner, M. W., Tilghman, S., & Varmus, H. (2014). Rescuing US biomedical research from its systemic flaws. *Proceedings of the National Academy of Sciences of the United States of America, 111*, 5773–5777.

Anderson, E. M., & Shannon, A. L. (1995). Towards a conceptualisation of mentoring. In T. Kerry, & A. Shelton (Eds.), *Issues in mentoring.* London: A.S. Routledge.

Anderson-Thompkins, S. (2016). *Nominations for the 2016 office of postdoctoral affairs mentor award.* [Online]. Available: http://research.unc.edu/postdoctoral-affairs/postdocs/npaw2016/.

Austin, A. E. (2002). Preparing the next generation of faculty: Graduate school as socialization to the academic career. *The Journal of Higher Education, 73*, 94–122.

Austin, A. E., & Mcdaniels, M. (2006). Preparing the professoriate of the future: Graduate student socialization for faculty roles. *Higher education: Handbook of theory and research.*

Barnes, B. B. (2009). The role of doctoral advisors: A look at advising from the advisors perspective. *Innovative Higher Education, 33,* 297–315.

Beach, A. L., Sorcinelli, M. D., Austin, A. E., & Rivard, J. K. (2016). *Faculty development in the age of evidence: Current practices, future imperatives.* Sterling, VA: Stylus Publishing.

Bertoline, G. (2015). *Trustees approve new guidelines for promotion and tenure to highlight mentoring.*

Bourdieu, P. (1986). *The forms of capital.* New York, NY: Greenwood Press.

Bozeman, B., & Feeney, M. K. (2007). Toward a useful theory of mentoring – a conceptual analysis and critique. *Administration & Society, 39,* 719–739.

Byars-Winston, A. (2016). *Culturally aware mentoring: A new mentor training module.* [Online]. Available: https://nrmnet.net/culturally-aware-mentoring-a-new-mentor-training-module/.

Byars-Winston, A., Gutierrez, B., Topp, S., & Carnes, M. (2011). Integrating theory and practice to increase scientific workforce diversity: A framework for career development in graduate research training. *CBE Life Sciences Education, 10,* 357–367.

Carlone, H. B., & Johnson, A. (2007). Understanding the science experiences of successful women of color: Science identity as an analytic lens. *Journal of Research in Science Teaching, 44,* 1187–1218.

Carruthers, J. (1993). The principles and practices of mentoring. In B. J. Caldwell, & E. M. A. Carter (Eds.), *The return of the mentor: Strategies for workplace learning.* London: Falmer Press.

Chemers, M., Zurbriggen, E., Syed, M., Goza, B., & Bearman, S. (2011). The role of efficacy and identity in science career commitment among underrepresented minority students. *Journal of Social Issues, 67,* 469–491.

Cho, C. S., Ramanan, R. A., & Feldman, M. D. (2011). Defining the ideal qualities of mentorship: A qualitative analysis of the characteristics of outstanding mentors. *The American Journal of Medicine, 124,* 453–458.

Cohen, G. L., Garcia, J., Apfel, N., & Master, A. (2006). Reducing the racial achievement gap: A social-psychological intervention. *Science, 313,* 1307–1310.

Committee on Science, E, Public Policy; Institute of Medicine, National Academy of Sciences, & National Academy of Engineering. (2014). *The postdoctoral experience revisited. The postdoctoral experience revisited.* Washington (DC).

Crisp, G., Baker, V., Griffin, K., Lunsford, L. G., & Pifer, M. (2017). Mentoring undergraduate students. *Association for the Study of Higher Education, 43,* 7–103.

Davis, G. (2005). Doctors without orders: Highlights of the Sigma Xi postdoc survey. *American Scientist, 93*(Suppl.).

Davis, G. (2009). Improving the postdoctoral experience: An empirical approach. In R. B. Freeman, & D. L. Goroff (Eds.), *Science and engineering careers in the United States: An analysis of markets and employment.* University of Chicago Press.

Davis, C. S., Ginorio, A. B., Hollenshead, C. S., Lazarus, B. B., & Rayman, P. M. (1996). *The equity equation: Fostering the advancement of women in the sciences, mathematics, and engineering.* San Francisco: Jossey-Bass.

Dolan, E., & Johnson, D. (2009). Toward a holistic view of undergraduate research experiences: An exploratory study of impact on graduate/postdoctoral mentors. *Journal of Science Education and Technology, 18,* 487–500.

Ensher, E. A., Thomas, C., & Murphy, S. E. (2001). Comparison of traditional, step-ahead, and peer mentoring on protégés' support, satisfaction, and perceptions of career success: A social exchange perspective. *Journal of Business and Psychology, 15,* 419–438.

FASEB. (2003). *Individual development plan framework* [Online]. Available. http://www.faseb.org/portals/2/pdfs/opa/idp.pdf.

Felder, P. P., Stevenson, H. C., & Gasman, M. (2014). Understanding race in doctoral student socialization. *International Journal of Doctoral Studies, 9*, 21–42.

Fetzer, J. (2008). Roles and responsibilities of graduate students and post-docs. *Analytical and Bioanalytical Chemistry, 392*, 1251–1252.

Figueroa, T., & Hurtado, S. (2013). *Underrepresented racial and/or ethnic minority (URM) graduate students in STEM disciplines: A critical approach to understanding graduate school experiences and obstacles to degree progression.* Los Angeles, CA: Association for the Study of Higher Education/University of California, Los Angeles.

Fleming, M., House, S., Hanson, V. S., Yu, L., Garbutt, J., McGee, R., et al. (2013). The mentoring competency assessment: Validation of a new instrument to evaluate skills of research mentors. *Academic Medicine: Journal of the Association of American Medical Colleges, 88*, 1002–1008.

Fuhrmann, C. N. (2016). Enhancing graduate and postdoctoral education to create a sustainable biomedical workforce. *Human Gene Therapy, 27*, 871–879.

Fuhrmann, C. N., Halme, D. G., O'sullivan, P. S., & Lindstaedt, B. (2011). Improving graduate education to support a branching career pipeline: Recommendations based on a survey of doctoral students in the basic biomedical sciences. *CBE Life Sciences Education, 10*, 239–249.

Fuhrmann, C. N., Hobin, J. A., Lindstaedt, B., & Clifford, P. S. (2015). *Science*. My IDP [Online] Science Careers. Available: http://myidp.sciencecareers.org/.

Gibbs, K. D., Jr., Basson, J., Xierali, I. M., & Broniatowski, D. A. (2016). Decoupling of the minority PhD talent pool and assistant professor hiring in medical school basic science departments in the US. *eLife, 5*, e21393.

Gibbs, K. D., Jr., McGready, J., Bennett, J. C., & Griffin, K. (2014). Biomedical science Ph.D. career interest patterns by race/ethnicity and gender. *PLoS One, 9*, e114736.

Gibbs, K. D., Jr., Mcgready, J., & Griffin, K. (2015). Career development among American biomedical postdocs. *CBE Life Sciences Education [Electronic Resource], 14*, ar44.

Golde, C. M., & Dore, T. M. (2001). *At cross purposes: What the experiences of today's doctoral students reveal about doctoral education.* PEW Charitable Trust.

Golde, C. M., & Dore, T. M. (2005). The role of the department and discipline in doctoral student attrition: Lessons from four departments. *The Journal of Higher Education, 76*.

Gonzales, P. M., Blanton, H., & Williams, K. J. (2002). The effects of stereotype threat and doubleminority status on the test performance of Latino women. *Personality & Social Psychology Bulletin, 28*, 659–670.

Gonzalez, J. C. (2006). Academic socialization experiences of Latina doctoral students: A qualitative understanding of support systems that aid and challenges that hinder the process. *Journal of Hispanic Higher Education, 5*, 347–365.

Griffin, K., Gibbs, K. D., Jr., Bennett, J., Staples, C., & Robinson, T. (2015). "Respect me for my science": A Bordieuian analysis of women scientists' interactions with faculty and socialization into science. *Journal of Women and Minorities in Science and Engineering, 21*, 159–179.

Gruner, S., Easley, D., Giese, E., Healey, T., Lin, H., Lenetsky, M., et al. (2015). *Report of the college of arts and sciences committee on streamlining research administration.* Ithaca, NY: Cornell University.

Gutierrez, B., Kaatz, A., Chu, S., Ramirez, D., Samson-Samuel, C., & Carnes, M. (2014). "Fair play": A Videogame designed to address implicit race bias through active perspective taking. *Games for Health Journal, 3*, 371–378.

Haley, K., Jaeger, A. J., & Levin, J. S. (2014). The influence of cultural social identity on graduate student career choice. *Journal of College Student Development, 55*, 101–119.

Henderson, C., Dancy, M., & Niewiadomska-Bugaj, M. (2012). Use of research-based instructional strategies in introductory physics: Where do faculty leave the innovation-decision process? *Physical Review Physics Education Research, 8,* 1–15.

von Hippel, T., & von Hippel, C. (2015). To apply or not to apply: A survey analysis of grant writing costs and benefits. *PLoS One, 10.*

Homer, & Fitzgerald, R. (1990). *The Odyssey.* New York: Vintage Books.

Ibarra, R. A. (2001). *Beyond affirmative action: Reframing the context of higher education.* Madison, WI: University of Wisconsin Press.

Jaeger, A., Haley, K., Ampaw, F., & Levin, J. (2013). Understanding the career choice for under-represented minority doctoral students in science and engineering. *Journal of Women and Minorities in Science and Engineering, 19,* 1–16.

Janasz, D. E., & Sullivan, S. E. (2004). Multiple mentoring in academe: Developing the professorial network. *Journal of Vocational Behavior, 64,* 263–283.

Kram, K. E., & Isabella, L. A. (1985). Mentoring alternatives: The role of peer relationships in career development. *Academy of Management Journal, 28,* 110–132.

Kuhn, C., & Castano, Z. (2016). Boosting the career development of postdocs with a peer-to-peer mentor circles program. *Nature Biotechnology, 34,* 781–783.

Laudel, G., & Gläser, J. (2008). From apprentice to colleague: The metamorphosis of early career researchers. *Higher Education, 55,* 387–406.

Lee, S. P., McGee, R., Pfund, C., & Branchaw, J. (2015). Mentoring up: Learning to manage your mentoring relationships. In W. G (Ed.), *The mentoring Continuum: From graduate school through tenure.* Syracuse, NY: The Graduate School Press.

Lencioni, P. (2002). *The five dysfunctions of a team.* Jossey-Bass.

Lent, R., Brown, S., & Hackett, G. (1994). Toward a unifying social cognitive theory of career and academic interest, choice, and performance. *Journal of Vocational Behavior, 45,* 79–122.

Little, J. W. (1990). *The mentor phenomenon and the social organisation of teaching. Review of research in education.* Washington DC: American Educational Research Association.

Lovitts, B. E. (2001). *Leaving the ivory tower: The causes and consequences of departure from doctoral study.* Lanham, MD: Rowman and Littlefield.

MacLachlan, A. J. (2006). *Developing graduate students of color for the professoriate in science, technology, engineering, and mathematics.* Berkeley, CA: Center for Studies in Higher Education.

Manson, S. M. (2009). Personal journeys, professional paths: Persistence in navigating the crossroads of a research career. *American Journal of Public Health, 99*(Suppl. 1), S20–S25.

Mason, M., Goulden, M., & Frasch, K. (2009). Why graduate students reject the fast track: A study of thousands of doctoral students shows that they want balanced lives. *Academe, 95,* 11–16.

McDowell, G. S., Gunsalus, K. T., Mackellar, D. C., Mazzilli, S. A., Pai, V. P., Goodwin, P. R., et al. (2014). Shaping the future of research: A perspective from junior scientists. *F1000Res, 3,* 291.

McGee, R. (2016). Biomedical workforce diversity: The context for mentoring to develop talents and foster success within the 'pipeline'. *AIDS and Behavior, 20*(Suppl. 2), 231–237.

McGee, R., 2017. RE: Personal communication.

McGee, R., & Keller, J. L. (2007). Identifying future scientists: Predicting persistence into research training. *CBE Life Sciences Education [electronic Resource], 6,* 316–331.

Merton, R. K. (1968). The Matthew effect in science: The reward and communication systems of science are considered. *Science, 159,* 56–63.

Meyers, F. J., Mathur, A., Fuhrmann, C. N., O'brien, T. C., Wefes, I., Labosky, P. A., et al. (2016). The origin and implementation of the broadening experiences in scientific training programs: An NIH common fund initiative. *FASEB Journal, 30,* 507–514.

Bennett, L. M., Maraia, R., & Gadlin, H. (2014). The 'welcome letter': A useful tool for laboratories and teams. *Journal of Translational Medicine & Epidemiology, 2*(2).

Miller, J. M., & Feldman, M. P. (2015). Isolated in the lab: Examining dissatisfaction with postdoctoral appointments. *The Journal of Higher Education, 86,* 697–724.

Miyake, A., Kost-Smith, L. E., Finkelstein, N. D., Pollock, S. J., Cohen, G. L., & Ito, T. A. (2010). Reducing the gender achievement gap in college science: A classroom study of values affirmation. *Science, 330,* 1234–1237.

Muller, M. (2009). *How to be proactive in your mentoring relationships.* [Online] National Postdoctoral Association. Available: http://www.nationalpostdoc.org/?page=Proactive.

NPA. (2017). *Call for nominations: National postdoctoral association Garnett-Powers and Associates Inc mentor award.* [Online]. Available: http://engage.asbmb.org/2016/07/26/nominate-someone-for-the-2017-national-postdoctoral-association-award-for-mentoring/.

NRMN. National Research Mentoring Network [Online]. Available: https://nrmnet.net/.

Pfund, C., Branchaw, J., & Handelsman, J. (2015). *Entering mentoring* (2nd ed.). New York, NY: W.H. Freeman & Company.

Pfund, C., Byars-Winston, A., Branchaw, J., Hurtado, S., & Eagan, K. (2016). Defining attributes and metrics of effective research mentoring relationships. *AIDS and Behavior, 20,* 238.

Pfund, C., House, S. C., Asquith, P., Fleming, M. F., Buhr, K. A., Burnham, E. L., et al. (2014). Training mentors of clinical and translational research scholars: A randomized controlled trial. *Academic Medicine: Journal of the Association of American Medical Colleges, 89,* 774–782.

Pfund, C., House, S., Spencer, K., Asquith, P., Carney, P., Masters, K. S., et al. (2013). A research mentor training curriculum for clinical and translational researchers. *Clinical and Translational Science, 6,* 26–33.

Phillips, C. J. (2010). *Let's talk! Expanding dialogue in the postdoctoral community towards broadening participation in the social, behavioral, and economic (SBE) sciences.* Washington, DC: The Postdoctoral Experience in the SBE Sciences.

Pickett, C. L., Corb, B. W., Matthews, C. R., Sundquist, W. I., & Berg, J. M. (2015). Toward a sustainable biomedical research enterprise: Finding consensus and implementing recommendations. *Proceedings of the National Academy of Sciences of the United States of America, 112,* 10832–10836.

Roberts, A. (1999). *Homer's mentor: Duties fulfilled or misconstrued?* (History of education).

Rothschild, M. L. (1999). Carrots, sticks, and promises: A conceptual framework for the management of public health and social issue behaviors. *Journal of Marketing, 63,* 24–37.

Rowe, M. P. (1990). Barriers to equality: The power of subtle discrimination to maintain unequal opportunity. *Employee Responsibilities and Rights Journal, 3,* 153–163.

Sambunjak, D., Straus, S. E., & Marusic, A. (2006). Mentoring in academic medicine: A systematic review. *JAMA, 296,* 1103–1115.

Santucci, A. K., Lingler, J. H., Schmidt, K. L., & Nolan, B. A. D. (2008). *Peer mentored research development meeting: A model for successful peer mentoring among junior level researchers* (vol. 32. Academic Psychiatry, 493.

Sauermann, H., & Roach, M. (2012). Science PhD career preferences: Levels, changes, and advisor encouragement. *PLoS One, 7.*

Scaffidi, A. K., & Berman, J. E. (2011). A positive postdoctoral experience is related to quality supervision and career mentoring, collaborations, networking and a nurturing research environment. *The Journal of Higher Education, 62,* 685.

Shea, G. (1997). *Mentoring* (rev. ed.). Menlo Park, CA: Crisp Publications.

Smith, M. H., Beaulieu, L. J., & Seraphine, A. (1995). Social capital, place of residence and college attendance. *Rural Sociology, 60*, 363–381.

Solórzano, D. G. (1993). *The road to the doctorate for California's Chicanas and Chicanos: A study of Ford foundation minority fellows* (Report to the California Policy Seminar, a joint program of the University of California and state government).

Steele, C. M. (1997). A threat in the air: How stereotypes shape intellectual identity and performance. *The American Psychologist, 52*, 613–629.

Steele, C. M., & Aronson, J. (1995). Stereotype threat and the intellectual test performance of African Americans. *Journal of Personality and Social Psychology, 69*, 797–811.

Straus, S. E., Johnson, M. O., Marquez, C., & Feldman, M. D. (2013). Characteristics of successful and failed mentoring relationships: A qualitative study across two academic health centers. (88). Academic Medicine.

Su, X. (2011). Postdoctoral training, departmental prestige and scientists' research productivity. *The Journal of Technology Transfer, 36*, 275–291.

Thomas, K. M., Willis, L. A., & Davis, J. (2007). Mentoring minority graduate students: Issues and strategies for institutions, faculty, and students. *Equal Opportunities International, 26*, 178–192.

Tinto, V. (1993). *Leaving College: Rethinking the Causes and Cures of student Attrition*. Chicago, IL: University of Chicago Press.

Trevelyan, R. (2001). The paradox of autonomy: A case of the academic research scientists. *Human Relations, 54*, 495–525.

Tyler, T. R., & Blader, S. L. (2003). The group engagement model: Procedural justice, social identity, and cooperative behavior. *Personality and Social Psychology Review, 7*, 349–361.

Wieman, C., Perkins, K., & Gibert, S. (2010). Transforming science education at large research universities: A case study in progress. *Change, 42*, 7–14.

Williams, D. A., Berger, J. B., & Mcclendon, S. A. (2005). *Toward a model of inclusive excellence and change in postsecondary institutions*. Washington, DC: Association of American Colleges and Universities.

Williams, S. N., Thakore, B. K., & McGee, R. (2015). Coaching to augment mentoring to achieve faculty diversity: A randomized controlled trial. *Academic Medicine: Journal of the Association of American Medical Colleges, 91*(8).

Chapter 6

Career Coherence, Agency, and the Postdoctoral Scholar

Karen J. Haley[1], Tara D. Hudson[2], Audrey J. Jaeger[3]
[1]Portland State University, Portland, OR, United States; [2]Kent State University, Kent, OH, United States; [3]North Carolina State University, Raleigh, NC, United States

Chapter Outline

Conceptual Framework: Agency and Career Coherence 122
Mixed Methods Research Design 124
    Instruments and Participants 124
    Coding and Analysis 126
    Researchers and Limitations 126
STEM Postdocs Act to Realize Their Career Goals 127
    Survey Findings: STEM Postdocs as Agents of Their Own Career Development 127
    Learning and Acting: The Career Coherence of STEM Postdocs 127

    Career Literacy 129
    Career Context 130
    Career Gumption 132
    Career Integrity 134
Three Primary Patterns of Agency and Career Goal Clarity 135
    High Agency–High Goals 135
    High Agency–Low Goals 136
    Low Agency–Low Goals 137
Discussion and Implications 138
Conclusions 141
Acknowledgments 141
References 141

By its design, a postdoctoral appointment should provide a recent doctoral graduate with directed, advanced training needed for them to become an independent scholar in a particular field: It is "a temporary period of mentored research and/or scholarly training for the purpose of acquiring the professional skills needed to pursue a career path of his or her choosing" (National Postdoctoral Association [NPA], n.d., para. 1). In science, technology, engineering, and mathematics (STEM) fields, a postdoctoral position is often considered a prerequisite for a faculty career (National Science Board, 2016; Nerad & Cerny, 1999), including 54% in biological, agricultural, and environmental life sciences, 45% in the physical sciences, and 19% in engineering (National Science Board, 2016). Postdocs are less common in fields outside STEM although their numbers are growing as noted in chapter one (e.g., in 2013, 18% of those in psychology and 6% in all

The Postdoc Landscape. http://dx.doi.org/10.1016/B978-0-12-813169-5.00006-9

other social sciences held postdoc positions within 3 years of completing their doctorates National Science Board, 2016).

However, as the number of postdocs in all fields in the United States and worldwide continues to increase, the level of dissatisfaction with these positions remains a consistent theme in the literature (e.g., Camacho & Rhoads, 2015; Miller & Feldman, 2015; Su, 2014; van der Weijden, Teelken, de Boer, & Drost, 2016; Wei, Levin, & Sabik, 2012). Postdocs' access to career development opportunities and the quality of those scant opportunities vary within and across institutions (Camacho & Rhoads, 2015; National Academies, 2014; Puljak & Sharif, 2009). This chapter explores the experiences of STEM postdocs in the United States in advancing their career goals. The goal is to provide insight into the most effective ways to support postdocs' career advancement.

CONCEPTUAL FRAMEWORK: AGENCY AND CAREER COHERENCE

This work is framed by agency theory, using O'Meara, Campbell & Terosky's (2011) definition of agency as an attribute of an academic member that is evidenced by his/her assuming strategic perspectives and/or taking strategic actions toward goals that matter. Using this conceptualization, agency includes two areas: perspective taking (or making meaning of situations and contexts in ways that advance personal goals) and the actions taken to pursue goals in a given situation (O'Meara et al., 2011). Neumann and Pereira (2009) suggest that agency is enacted in relationship to something, and individuals can display agency in one area but not in another. For example, a postdoc may assume agency in the lab in which he or she works but not feels agency in terms of making important career decisions to prepare for the next position; therefore the capacity may be "project specific," or only enacted in certain contexts. Utilizing agency theory to frame the participants' experiences, we were interested in the perspectives assumed and actions taken by postdoctoral scholars in pursuit of their next career decision and career goals (Archer, 2003).

While agency provides the overarching concept for this study, Magnusson and Redekopp's (2011) model of coherent career practice provides a detailed structure to delve into the project-specific context of career goals. The model integrates elements common to varied theories of career development and is comprised of four components: (1) career literacy, (2) career context, (3) career gumption, and (4) career integrity. *Career literacy* refers to an awareness of what is necessary to pursue a desired career; specifically, it includes "a progressively acquired set of skills, knowledge, and attitudes that are related to the acquisition, understanding, and application of information needed to manage one's own career development" (Magnusson & Redekopp, 2011, p. 176). *Career context* pertains to an individual's perceptions of career possibilities situated within their immediate environment and, more abstractly, in the world at large. Magnusson and Redekopp note, "Career context creates the career environment in which one's choices are made" (p. 177). Potentially, literacy and context have elements of both agentic actions (acquiring skills or seeking out contexts) and agentic perspectives (perceiving their choices). *Career gumption* refers to "the energy, momentum, motivation, or desire to engage in career development" and "ties together motivation, optimism, hope, and self-efficacy" (Magnusson & Redekopp, 2011, p. 177). Career gumption relates to both the action and perspective components of agency as defined by O'Meara et al. (2011), who explain that exercising gumption necessitates "assuming strategic perspectives and/or taking strategic actions toward [career-related] goals" (O'Meara et al., 2011, p. 157). Finally, *career integrity* represents "the extent to which individuals made personally satisfying choices within the available career options given their desire to balance career and personal aspirations and personal values" (McAlpine, 2016, p. 10). An individual achieves career integrity by achieving "a meaningful balance between personal, social, economic, and community factors" as well as "congruence between one's identity and the roles one plays" (Magnusson & Redekopp, 2011, p. 177). When an individual feels he or she has achieved a desired balance and is satisfied with career choices he or she has made given his or her options, career integrity is present.

Together, these four elements of career coherence comprise the process of career integration. Magnusson and Redekopp (2011) explain that because these four elements constitute a dynamic process, "change in any one part of the system creates change in all other parts of the system" (p. 178). For example, a loss of career integrity may reduce an individual's career gumption and vice versa. In addition, struggles or "blockages" in any of the four elements or in integration can impede career growth and progress.

The coherent career practice model was developed for practice in a counseling context, although a wider application, especially in tandem with agency theory, is a good theoretical match. There has been limited attention by researchers to the ways individuals utilize individual agency to overcome struggles that may impede meeting career goals; however, McAlpine (2016) drew on the concept of career integrity in her study of PhD students' paths toward nonacademic careers and found that participants made career choices that allowed them to maximize their career integrity within the constraints posed by the choices available to them and their personal values.

Using these two frameworks together, we connect postdoc agentic perspectives and actions to the framework of coherent career practice. *Career literacy* and *career context* are elements of the career coherence framework related to what an individual knows about the desired career and how he or she perceives his or her ability to attain it. *Career integrity*, driven by an individual's desires, needs, and priorities, may be shaped by career literacy and context and, subsequently, may drive career gumption. These elements are all intertwined and as Magnusson and Redekopp (2011) suggest, "change in any one part of the system creates change in all other parts of the system" (p. 178). Furthermore, these elements are all influenced by an individual's agentic perspectives and actions. Since many postdoctoral researchers report facing career uncertainty and a dearth of available academic jobs (and, to a large extent, industry jobs as well), it is important for institutions that are increasingly relying on postdocs to support the waning faculty to know the impact of the position on their employees' future career options. It is essential that policy addresses how to best serve this population in terms of their career advancement, especially since position in and of itself is largely meant to be an element of career development for the postdoc.

The current study explores agency within the framework of career coherence for postdoctoral scholars in STEM fields to better understand how postdocs think about and take action to realize their career goals. In addition, because Magnusson and Redekopp (2011) note that their model of coherent career practice has room for development, our study also serves the purpose of applying the model to a new population and expanding it to intentionally incorporate the concept of agency.

MIXED METHODS RESEARCH DESIGN

This research was part of a large mixed methods design study. A mixed methods approach was warranted for several reasons. First, given agency theory has not been used with postdoctoral scholars, we wanted to understand the presence and kinds of agentic perspectives and actions in the postdoc experience. A survey with items focused on agentic perspectives and actions accomplished that purpose. However, we also wanted to understand agentic perspectives and actions within the everyday career-related experiences of postdocs. We wanted a richer set of concrete examples than a survey could provide. We also wanted to understand the relationship between the postdocs' expressions of agency and their career decision advancement processes. This was best accomplished through qualitative interviews.

A mixed methods study design contains both quantitative and qualitative elements, which can be sequential or concurrent, equal weight or prioritized (Cresewell & Plano Clark, 2011). The survey in this study was the initial contextual component that set the stage for the individual interviews. Therefore, the qualitative portion will be the focus of this chapter and will have the explanatory role (Cresewell & Plano Clark, 2011).

Instruments and Participants

The online survey was developed by a team of researchers based on current research on agentic perspectives and agentic action in relation to career goals. The survey was sent to all postdocs at the study institution, a major research university located in the southeastern United States. The postdoctoral development office provided names and email addresses of all 341 postdocs employed by the institution. Of those 341, 172 surveys were returned for a 50% return rate.

The survey offered us the opportunity to engage with the postdocs in a way that connected them to the project before asking if they would be interested in participating in an in-person interview; 72 respondents expressed interest. From among these 72 respondents, we arranged interviews with postdocs who represented a diverse mix of discipline, gender, and race. This criterion sampling technique (Mertens, 2010) to recruit, identify, and select participants yielded 19 interviews.

While the 19 participants did not represent all disciplines at the study site, they represented 17 different departments. Additionally, seven of the participants were persons of color, three identified as international, and seven were women, providing a diversity of postdoc experiences (see Table 6.1). While we

TABLE 6.1 Participant Self-Identified Academic Discipline, Race/Ethnicity/Status, and Gender

Academic Discipline	Race/Ethnicity	Gender
Astrophysics	International	Male
Biochemistry	White	Female
Biology/Ecology	White	Female
Biology/Plant Pathology	Latino/Hispanic	Female
Biology/Science Education	White	Female
Chemical Engineering	Latino/Hispanic	Male
Crop Science/Plant Biology	White	Male
Entomology	White	Male
Experimental Nuclear Physics	International	Male
Fiber Polymer Science	International	Male
Food Safety/Nutritional Science	White	Female
Forestry and Environmental Resources	White	Female
Genetics	African–American	Male
Geochemistry	Latino/Hispanic	Male
Mathematics Education	Latino/Hispanic	Female
Microbiology	Latino/Hispanic	Male
Poultry Science	White	Male
Statistics	White	Male
Toxicology/Genetics	Latino/Hispanic	Male

identified gender, race, and nationality to demonstrate the level of participant diversity, we minimize our use of these identities in reporting our findings to protect participant confidentiality.

The interviews lasted between 60 and 90 min. The semistructured interview format allowed flexibility for participants to guide the direction of the interview (Knox & Burkard, 2009). The interview questions focused on the actions participants had taken and the perspectives they had adopted to pursue their career goals as well as aspects of their department environment and discipline that seemed to influence their sense of agency. Face-to-face interviews enabled researchers to detect social cues and nonverbal communication, elements that provided deeper insights into the participants' experiences (Opdenakker, 2006). With participant permission, all interviews were audio recorded in preparation for transcription. The authors would like to extend their gratitude to the initial team of doctoral students including Jingjing Zhang, Ashley Grantham, Allison Mitchall, and Alessandra Dinin lead by their faculty advisor, Audrey Jaeger, one of the authors of this chapter, who gathered the data for this research project.

Coding and Analysis

The 19 interview transcripts were read and annotated multiple times until the researchers had a strong grasp of each participant's perceptions and experiences (Creswell, 2007). The researchers created a codebook using keywords drawn from O'Meara et al. (2011) agency theory (agentic perspective and agentic action) and Magnusson and Redekopp's (2011) model of coherent career practice (literacy, context, gumption, and integrity). We coded the transcripts for evidence of each of the six codes pertaining to concepts within our two conceptual frameworks as well as for interaction between the codes (DeCuir-Gunby, Marshall and McCulloch, 2011). In this process, the codes were defined, refined, and verified to ensure interresearcher reliability (Marshall & Rossman, 2011).

We created a profile for each of the 19 interview participants, which summarized data pertaining to each of the four career coherence codes and provided an overall assessment of the participant's career coherence. Additionally, we used magnitude coding (Saldaña, 2016) to classify each participant as low, medium, or high along each of the four career coherence codes. Based on the magnitude coding and participant profiles, we then positioned each participant within a four-quadrant matrix (Saldaña, 2016) representing two dimensions: agency (low to high) and career goal clarity (low to high). This approach allowed us to classify participants into three patterns.

Researchers and Limitations

The data were collected by a large research team of faculty and doctoral students. The researchers for this article reviewed the data and completed the qualitative analysis. This process separated the data collection and analysis, which could be both an advantage and a limitation. One researcher who helped collect all forms of data provided a context and continuity of analysis, and the two additional researchers on this article offered alternative insights, but they may have missed some of the nuances found during the interview process. Another limitation of the study is the focus on STEM postdocs, and it is unknown how these findings could be generalized to explain postdoc experiences in the humanities. Overall, the study integrity was affirmed through multiple perspectives and triangulation of the data.

STEM POSTDOCS ACT TO REALIZE THEIR CAREER GOALS

This study sought to understand and explain how STEM postdocs define and take action to realize their professional career goals. To explain the process we used to answer this question, we first present a summary of the results from the agency-related survey items. We then turn to the findings from our interviews, which provide deeper insight into our research question. The analysis begins with a summary of our descriptive findings regarding career coherence of postdocs and closes with an analysis of the three primary postdoc agency/career coherence patterns.

Survey Findings: STEM Postdocs as Agents of Their Own Career Development

The initial survey gathered data from 172 postdocs who gave an overview of general patterns of agentic perspectives and agentic actions. Table 6.2 presents the survey question, whether it is categorized as perspective (P) or action (A), the number of responses (N), the mean (AVE), and the standard deviation (SD). The range for the Likert-type scores was 1 (low) to 5 (high). Mean scores for the positively worded questions ranged from 3.84 to 4.19, and mean scores for the negatively worded questions ranged from 2.34 to 2.88.

The self-reported scores were consistently above 4.0 for the positively worded questions and lower than 3.0 for the negatively worded questions, which reflect positive agentic perspective and action. The survey results provided a baseline for the interview process, as we could then delve deeper into specific examples of agency.

Learning and Acting: The Career Coherence of STEM Postdocs

The interview data were rich in examples of agentic perspectives and agentic actions. Overall, we saw a wider range of apparent agency than was shown in the survey results, thereby meeting our goal of using research design as a way of delving deeper into the lived experience of postdocs. Agency is the overriding concept of the study, and the exercise of agency requires a specific context. Therefore, we locate agency within the context of career choices and goals using the four components of the career coherence model (literacy, context, gumption, and integrity) to provide the structure for presenting the data (Magnusson & Redekopp, 2011).

TABLE 6.2 Results From Survey of Postdoctoral Scholars in STEM

A/P	Question	N	AVE	SD
A	I have been strategic in achieving my career goals.	171	3.84	0.81
A	I have intentionally made choices to focus my career goals in ways that are personally meaningful to me.	172	4.08	0.75
A	I seize opportunities when they are presented to me to advance my career goals.	172	4.13	0.77
P	I feel stuck in my ability to advance toward my career goals.	170	2.88	1.02
P	In general, I feel that I have little control over whether I advance my career goals.	172	2.65	1.06
A	If I lack something (e.g., a skill or specific knowledge) that I need to obtain my career goals, I take steps to obtain it.	172	4.03	0.78
P	My adviser (rather than I) controls whether I will achieve my career goals.	172	2.34	1.01
P	I know how to conduct research to achieve my career goals.	172	4.06	0.64
A	When I face a setback, I take strategic steps to overcome the barrier to move ahead.	171	4.05	0.62
P	When I face a setback, I view it as a temporary roadblock that I can overcome.	172	3.99	0.73
P	I can acquire all of the knowledge and skills that I need to be successful in my career goals as long as I work hard.	172	3.74	1.02
P	I ask for help when I need it.	172	3.95	0.76
P	I view critical feedback on my work as a way to grow.	171	4.19	0.69
P	Rather than seeing one path, I see there are multiple paths to be successful in achieving my career goals.	172	4.01	0.83

A, action; *AVE*, mean; *N*, number of responses; *P*, perspective; *SD*, standard deviation.

Career Literacy

As defined by Magnusson and Redekopp (2011) in their model of coherent career practice, career literacy refers to an awareness of what is necessary to pursue a desired career and includes "skills, knowledge, and attitudes that are related to the acquisition, understanding, and application of information needed to manage one's own career development" (p. 176). The primary form of career literacy evident among participants was recognition of the skills and knowledge they needed to develop relative to the specific career they sought, whether as a faculty member, in a government laboratory, or in industry. Participants who demonstrated higher levels of career literacy were reflective when identifying the specific skills and knowledge necessary for attaining their goals; they also sought ways to gain those by pursuing specific opportunities within their postdocs, often demonstrating career gumption in the process. One participant exemplified this reflective process:

> *I think about the job that I want to get, which would be either an academic position or a government position, and the sorts of skill sets that they would want in that type of person. ... So when I think about getting those jobs, I think about, "Where am I the strongest? And where am I the weakest?" ... the self-talk in terms of that is often like reevaluating what I've done and what I need to do in order to get where I want to go.*

Other participants discussed enhancing their career literacy by building their professional networks or augmenting their skill in teaching or specific laboratory techniques. Some participants also discussed "testing the waters" to explore different career possibilities and to determine the kinds of experiences they found most and least fulfilling (with an eye toward future career integrity). Some participants worked within the parameters of their postdoc appointments to enhance their career literacy, while others had to pursue professional development opportunities beyond their postdoc responsibilities.

Most of the postdocs in our study exhibited moderate to high levels of literacy related to the careers they intended to pursue; only three demonstrated low levels of career literacy. One participant with a high level of career literacy aspired to become a faculty member at a research institution and had a clear sense of what he needed to do to attain that position. After completing all 3 degrees at the same institution, he recognized that to be competitive for an academic position, he would need experience at a different institution, which is one reason he chose his current postdoc. His postdoc experiences have helped him better understand what a faculty member does:

> *I went to faculty meetings when my boss couldn't go and was on committees for things and all that kind of thing so I really got to see the ins and outs of what being a faculty member would be like and then make the decision of, 'Yeah, this is something I want to do.'*

He was aware of what he needed to do to be tenure-track position and kept a calendar for himself: "I've got my kind of my calendar right now is purely based on tenure ... Here's what I need to be doing on a year to year basis and by mid-tenure and tenure review." However, he expressed some uncertainty that he might not be doing everything he needs to be doing and wished he had more feedback to confirm that he is on the right track toward his faculty career.

The experiences of the previous participant contrast with those of another postdoc who also expressed a desire to become a faculty member yet exhibited a low level of career literacy. He took a passive approach to constructing his career path: "I guess kind of the default was to try to go into academics and I never really strayed from that, but I never really knew until my advisor said anything that that was going to be the possibility, I guess." Like the previous participant, he attended faculty meetings at the invitation of his advisor, which allowed him to "get perspective on what goes on inside that room." However, unlike the previous participant, he seemed content to allow his advisor to direct his career literacy rather than taking agentic action to enhance it on his own. He commented that one of his primary reasons for wanting to pursue a career as a faculty member was because he found his field of statistics interesting; he seemed unaware that an interest in the field, while necessary, is insufficient for success in a faculty career.

Similarly, one participant who sought a career in industry seemed to possess limited knowledge about what it would take to pursue the industry career he desires, and instead blamed the nature of the pharmaceutical industry for why he has been unsuccessful in obtaining an industry position, rather than reflecting on his own lack of relevant skills: "I'm confident I can learn [analytical techniques]; however, it doesn't matter how many times I tell them this, they're not willing to make the effort to let me do that. ... I'm not going to get hired for it."

Career Context

Career context pertains to an individual's perceptions of career possibilities situated within their immediate environment and, more abstractly, in the world at large: It "creates the career environment in which one's choices are made" (Magnusson & Redekopp, 2011, p. 177). The participants in our study identified a range of contextual factors influencing their perceived career options. Those seeking faculty careers frequently discussed the lack of available faculty positions and stiff competition: "The statistic on the numbers of PhDs we create and the number of available jobs, just, you can't expect everyone who graduates to get a job in faculty." Salaries were a relevant contextual factor for those choosing between careers in industry or academia, as was the perception expressed by several participants that a faculty career "means literally groveling for money to be able to do science." However, a few participants felt that an academic career offered more intellectual freedom than industry, where they believed that "you would basically do the work that someone else has designed." As one postdoc noted,

If I go into industry, you know, it's 9 to 5, you have to check in and ... there's set protocols on how to do things, whereas here [in academia] I can try different things and have just the ability to do what I want and how I want to do it.

Work–life balance issues were another salient factor for both male and female postdocs who sought faculty careers and had or planned to have a family. Several participants discussed being deterred from seeking faculty positions because of the long hours expected in the laboratory and the difficulty of achieving work–life balance as a faculty member. A postdoc in entomology commented,

For a long time … the goal was to get a faculty position. And then once going through that process and becoming more aware of what that involves, the amount of grant work particularly and the competitiveness of it, the number of hours that most people work and my desire to balance family life, I've decided that that may not be my only goal.

This same participant also discussed avoiding a specific type of laboratory that he felt would be incompatible with being a husband and father:

I've avoided working in molecular-based labs because in many cases those types of labs have a lot of demand on the workload. You're expecting long hours, experiments running all the time. … I have a family. I have a young child that I want to spend time with.

A few participants discussed how their career plans have changed over time due to perceived job market constraints. A postdoc noted, "When I started my PhD, the way that I thought about my job opportunities was very different than how I think about it now, certainly because a lot of places are on hiring freezes now." She had originally planned to work for the EPA, but "the way that the funding situation has gone for the EPA, I'm like, 'Well, maybe not so much.'"

A few female participants discussed their identity as women in science as a factor shaping their career contexts. One woman in nutritional science discussed how her identity as a woman "definitely impacts the way in which I go about getting, or thinking about how I'd like to get, my future jobs" because

You walk a fine line of wanting to be able to be noticed for your accomplishments but also, as a lady in the South, you don't want to brag about yourself, but you need to be assertive. So being sort of introverted and quiet to start off but then also trying to make sure that people notice when I have done a good job so that I can progress in my career.

Additionally, some participants discussed geographic considerations affecting their perceived career options, such as wanting (or not wanting) to relocate to a specific region, or, for international postdocs, limited career opportunities within their home countries. One participant discussed her choice to seek jobs outside the United States "because in some places, like Switzerland, they don't have the kinds of issues that we do at the moment."

Nearly three-quarters of participants exhibited moderate to high awareness of career context. These participants identified a wide range of contextual factors that shaped their perceived career options and were reflective about how their individual contexts influenced their career decision-making. For example, a postdoc in mathematics education discussed how her experiences teaching in graduate school shaped her career goal and the types of positions in which she is interested in pursuing after completing her postdoc:

> *"The experiences that I had teaching, I really enjoyed probably more so than the research part. And so I think that sort of helped me figure out that I don't necessarily want an academic position at a tier one research school but would prefer a smaller teaching school."*

In contrast, the five participants with low awareness of career context provided an incomplete assessment of the context in which they were making their career decisions. For example, one participant mentioned his desire "to interact with regular people" (i.e., not academics) as a motivation to choose a career located away from the region in which he currently lived, which has one of the highest concentrations of doctorate holders in the United States.

Levels of career context often correlated to levels of career literacy, as gaining (or lacking) information about the skills and knowledge needed to pursue one's desired career usually entailed learning more about the context for that career. One participant, a postdoc in genetics, "always wanted to be a faculty member," yet had also started to consider career in industry after "coming into the postdoc and just seeing what the market what looks like [for faculty]." However, he felt he lacked understanding regarding what a career in industry would entail (literacy) and did not have a sense of what his industry options might be (context).

Career Gumption

Career gumption refers to the action or energy put toward enacting career goals (Magnusson & Redekopp, 2011) and connects to both the action and perspective components of agency (O'Meara et al., 2011). The participants demonstrated a wide variety of activities that represent gumption. A few postdocs showed career gumption in seeking and attaining their postdoc position that indicated clear intentions in making decisions about their specific postdoc position and future career. Several participants exhibited high levels of gumption evidenced by their modifying or expanding their postdoc position to gain skills designed to meet their future career goals. One participant explicitly connected her goals: "So a lot of the goals that I set out for my postdoc are very geared towards the two types of positions that I want to do." Another gave a specific example: "When I started my postdoc, [I would] just have conversations with my advisor about my desire to have more teaching experience and working with her to be able to incorporate some of those opportunities into my postdoc experience."

Many postdocs took advantage of opportunities outside of the postdoc position to advance their career goals, either because they were not supported within the role or the role had limited opportunities. Two participants went beyond expectations of their advisor to find opportunities to increase their own skills.

One took a course on campus that is related to her work but outside of her department, and another said, "I go to these teaching seminars and workshops and take a vested interest in actually learning about how to be a better teacher and becoming one." Finally, while one participant took advantage of opportunities outside of her postdoc position, "it was always with the caveat [from her PI] that I did not let it detract from my work."

Networking, both as a skill and as a larger sphere of influence, was mentioned by several participants. They were very clear that networking was connected to their future career opportunities. While some noted that the act of networking was not comfortable, it was worth the effort, as noted by one participant who was willing to push herself outside of her comfort zone to ensure her career development:

I feel like a lot of getting it, the job that you want, is who you know, and in order to know those people, you have to be able to network. And since it's a little tougher for me to just be able to approach people and get to know them without some sort of premise behind it, I think I have to work extra hard at making sure that I put myself out there and in the right places at the right times.

In addition, the connection between networking and finding opportunities to demonstrate their skills in leadership and collaboration was evident. For example, "Lately, I've been kind of getting quite involved and getting my name out there and taking on projects, and doing some kind of collaboration meetings, leading them and sharing them and stuff." Building a strong and diverse professional network is an important priority and requires self-agency to engage with others.

Several of the postdocs were at the stage of applying for their next positions. One participant noted that after "a couple of job interviews that didn't necessarily go the way I had hoped, I was able to sort of take those experiences and work with my advisor to create some different opportunities in my postdoc that gave me more of the experience that I wanted, that compared to maybe when I first started my postdoc." The first time one participant applied for a tenure-track position, he did not know what to expect. While it was not a successful search, he got some great advice from a colleague: "You have to stop thinking like a postdoc. You're here asking for a job, but as a postdoc … think as a professor." These participants learned from their first application experiences and either went back to shore up their experiences or change their attitudes.

Almost three-quarters demonstrated moderate to high levels of gumption as described above. There were four who showed gumption in obtaining their current postdoc appointment and getting experiences within the postdoc role—increasing skills, gaining experience, and publishing—but they did not clearly tie those experiences to their future career goals. Finally, there were three participants who did not express gumption in their interviews or for whom it was coded as very low. These participants exhibited low levels of energy or apathy toward career goals. One participant said, "I haven't taken advantage of the opportunities to go and network and do things." Another knew he should be learning to write an effective proposal, but perceived it as a waste of time despite knowing that it was not. In a way, these postdocs knew what they should do, but did not make the effort.

Career Integrity

Career integrity represents "the extent to which individuals made personally satisfying choices within the available career options given their desire to balance career and personal aspirations and personal values" (McAlpine, 2016, p. 10). Almost all participants made a case for their decisions about career fit. Even when their goals were unclear, they still were making connections between what they wanted to do (tasks and roles) and their career options.

The participants who demonstrated higher levels of career integrity varied in how they identified these aspects. Several specifically named their families as an important factor in their career decision-making process. They wanted to balance their families' needs while achieving a work–life balance that allows them to be available for their kids and spouses. Being a parent and the desire to not relocate her family had shaped the career options for one participant—she did not see her family as a constraint so much as a factor to help her achieve career integrity—and she had the luxury of taking her time to find the right position. One participant noted,

> It's also how can I best provide for my family with the skills that I've got and also what my wife wants out of her life and what I want for my kid? Those have all come together to influence my career goals.

Another participant had a unique situation as her husband was in the military and moved every 3 years; she stated, "I always focus my decisions on what will be best for my children and also where I'll be next because my husband [gets moved] usually every 3 years."

Those who demonstrated career integrity also talked about intertwining identity and career fit as they wanted current positions that suited them based on the tasks/roles of their future positions, which was frequently driven by their career identity—as scientists, researchers, or teachers. Two identified as scientists or researchers, which guided their perceptions of career fit. For example, one said, "So laboratory work, I love it. I love the research and the science behind it, and so I'd like to continue in a career that has some aspect of that."

A few of the postdocs thought of themselves as teachers. One said she had been intentional about working with her advisor to ensure her postdoc experiences were aligned with her career goal of teaching, and another valued research only as it helped teach students: "Teaching is, obviously, my main focus; I enjoy research as a means to teach better, and to incorporate my students in their own learning journey." Others wanted a combination of teaching and research. One example of clearly connecting with both roles is one participant who really wanted to teach; however, she said, "I don't want to be in a place where teaching is so much the emphasis that research is just something you do in your extra time, because the thing that really gets me going in the morning is research."

While most participants exhibited moderate to high levels of career integrity, there were four who appeared to have career integrity without reflection on their choices, meaning they explicitly demonstrated their career fit but did not connect it to their identity, life balance, or values. One participant felt that a tenure-track position was a good fit because he liked to lecture and liked research. Another participant valued the application of research and sought a position in industry and hoped to later seek a teaching position that valued application. Several participants placed themselves firmly in the role of scientist outside the bounds of a particular position, which implied career integrity to the role but not necessarily to a career. Finally, there were three participants who did not address any aspect of career integrity during the interview.

Three Primary Patterns of Agency and Career Goal Clarity

In our final analysis, we found that the participants could be grouped into categories based on the clarity of their career goals. Those with clear career goals fell into two subgroups: (1) those who knew their career options in a broad sense (alternatives) or (2) those who were focused on one path (tunnel vision). A third group of participants held fairly clear goals yet were unsure if those goals were attainable. Finally, there were several participants who had unclear goals or were questioning their goals.

We also found that the postdocs in our study exhibited varied levels of agency. Many discussed taking agentic action and/or holding agentic perspectives related to their career goals, but the extent, or level, of agency they demonstrated varied, and a few participants seemed to have low levels of career-related agency. These insights led us to position each participant within a matrix (Saldaña, 2016) representing two dimensions: agency (low to high) and career goal clarity (low to high), thereby combining the coherent career practice (Magnusson & Redekopp, 2011) and agency (O'Meara et al., 2011, 2014) frameworks that guided our study. This approach allowed us to classify participants into three patterns based on agency and career goal clarity. Only one participant fell outside the patterns presented below.

High Agency–High Goals

The postdocs occupying this quadrant (n = 10) demonstrated high levels of career-related agency and clear, well-defined career goals. These participants engaged in agentic action to learn more about their career options, ensure their own professional development, and position themselves to achieve integrity in their future careers. They recognized that, ultimately, they were the only ones responsible for ensuring their future success. Within this group, if they were unable to attain satisfactory career development within their postdoc appointments, they sought opportunities to build their skills, develop their interests, and establish professional networks on their own outside of their current positions. These postdocs were reflective and tended to engage in agentic self-talk (i.e., perspectives regarding pursuit of career goals; see Jaeger et al., 2014; O'Meara et al., 2014), regularly assessing their career development relative to the skills and qualifications they believed they would need to attain their desired future careers and then taking action to address any perceived deficiencies.

While there was some variation in clarity and connection to current work, overall, the postdocs in this quadrant had well-defined career goals. For some, this meant that they had developed their career goal and stuck with the one vision (as we describe tunnel vision), while the rest had explored their options and made an informed choice. Both groups were striving toward a concrete career goal and knew the steps needed to get there.

Many in this group also had high levels of career integrity; they clearly connected who they were and what they valued to their career choices. Due to their high levels of agency and clear goals, they saw a clear path leading from their current postdoc positions to the career to which they aspired, and they knew what they needed to do in terms of their own professional development to successfully navigate that path, unlike their peers with lower agency and/or less clearly defined career goals. The postdocs in this group made choices that were their own, not someone else's ideas for them, and yet they included family needs and life–work balance as a part of the decision-making process.

High Agency–Low Goals

The postdocs occupying this quadrant (n = 4) demonstrated high levels of career-related agency despite not having clear career goals. Like the postdocs in the High Agency–High Goals pattern, these participants engaged in agentic action to ensure their professional development. They discussed exercising career gumption and engaging in reflective self-talk to enhance their career literacy and work toward integrity in their current postdoc positions and future careers.

Despite their high levels of agency, however, these postdocs struggled to gain traction working toward their future careers due to unclear or unfocused goals. Some expressed reluctance at closing the door on any possible opportunities, in part, because they recognized that the odds were not in their favor, especially for a faculty career. Due to a lack of goal clarity, their career development was decoupled from their career goals; many took an "all of the above" approach to their career development, pursuing as many opportunities as possible without thinking critically about which might represent the best investment of their time and resources. They tended to have only moderate amounts of literacy regarding the different career options under consideration possibly because their diffuse career goals kept them from fully understanding what the path to a specific career requires. While these postdocs tended to have a clear sense of what they would require to achieve integrity in their future careers (e.g., maximum work hours per week or a specific balance between teaching and research), they did not have a clear understanding of whether they could realistically obtain these things within the careers they were considering or they were not prepared to abandon the path they were on despite recognizing that their requirements for integrity were incompatible with that career.

The defining characteristic of this group of postdocs is of working hard to move themselves along a path despite not knowing where exactly that path would lead them or recognizing that their present path would likely be dead end. They seemed to believe that as long as they continued to work hard, they would eventually stumble onto career success. However, until these postdocs define their career goals more clearly, their high levels of agency may ultimately prove futile in helping them attain career coherence.

Low Agency–Low Goals

Postdocs located within this quadrant (n=4) exhibited low agency and unclear career goals. Their academic career paths to date had been shaped more by circumstances (such as gaining or losing funding for their research) than by intentionality (demonstrating career gumption or a drive toward career integrity), and their future career paths lacked definition. Several of these postdocs discussed aiming toward a faculty career, yet they had not thought critically about this option and seemed instead to see it as the default path; they possessed limited knowledge about what a faculty career would entail (career literacy) or whether it would be a good fit for them (career integrity). The postdocs in this quadrant who had considered alternatives to a future faculty career (e.g., working in industry or a government lab) similarly lacked literacy about these options, including the specific requirements of these jobs and whether they would be satisfied with the work environment. Two of the postdocs in this quadrant had slightly more defined career goals than the other two, in that they could identify specific tasks they wished to do or avoid in their future careers (such as amounts of time they wanted to devote to research, teaching, or seeking funding), yet they had not investigated whether realizing their aspirations would be feasible.

Regardless of chosen career path, these postdocs largely relied on other people, most commonly their graduate school and/or postdoc advisors, for their career development and planning, and they rarely took agentic action to learn more about their career options or to position themselves for success in pursuing those careers. Those who had engaged in some level of career planning discussed only general plans, such as publishing more, rather than reflecting on whether their plans would help them to develop the specific skills or qualifications needed to become a competitive candidate for their intended career paths. Most of their information regarding career options was received secondhand from advisors, peers, or other colleagues within their fields, and they assimilated this information without question, leading them to hold some unfounded assumptions about their career options. Postdocs in this group tended to wait for career development opportunities to come to them rather than pursuing action on their own. While their low career literacy, to some extent, kept them unaware of how to effectively advance their career development, a couple of postdocs in this quadrant discussed choosing not to do what they knew they "should" be doing, such as attending professional development opportunities or networking at conferences. Two attributed blame for their lack of career progress or success to date on factors they cannot control, such as highly competitive job markets or a lack of available funding for research.

The low levels of agency among postdocs in this group may have caused them to have undefined career goals, or perhaps the reverse is true, that their lack of goal direction stymied their agency. Regardless, it is unlikely these postdocs will achieve career coherence without both enhancing their agency and clarifying their career goals.

DISCUSSION AND IMPLICATIONS

Our survey found that postdocs self-report holding agentic perspectives and taking agentic actions similar to the work on doctoral student agency (O'Meara et al., 2014). Based on the survey results, we wanted to more deeply explore how postdocs in STEM fields define and take action to realize their professional career goals. Our interviews with 19 participants provided examples of how agency functions in the context of advancing career goals and illuminated the case of those who exhibited low levels of career-related agency.

The coherent career model (Magnusson & Redekopp, 2011) was developed as a clinical tool, but it provided us a clear structure to understand postdoc career goals and agency. Our results are more detailed than the only other research to use the model. McAlpine's (2016) study of PhD students' paths toward nonacademic careers, which found that career choices were made to maximize their career integrity within the constraints posed by the choices available to them and their personal values. Utilizing all four of Magnusson and Redekopp's components allowed us to create the four quadrants (agency and career goals) and describe the range of postdoc experience in enacting agency within the context of their career goals.

Interview data confirmed that a majority of postdocs do have high agency (found in two of our three primary patterns). While the simple majority of participating postdocs (n = 10) fell within the High–High pattern, eight participants seemed to lack clearly defined career goals (those within the Low–High and Low–Low patterns). Indeed, the experiences of the postdocs in our study, especially the 16 who were aiming for or considering faculty careers, echoed van der Weijden et al. (2016) conclusion that "postdocs seem to be trapped between their own ambitions and a lack of academic career opportunities" (p. 25).

Our findings regarding patterns of agency and goal clarity suggest three practical implications for supporting the career development of STEM postdocs. We orient our practical implications toward the needs of each of the three patterns of agency and goal clarity identified among the postdocs participating in this research. We concur with the National Academies (2014) that "the postdoctoral experience itself should be refocused, with training and mentoring at its center"

(p. 69), and this research and the implications we draw from it point toward ways to ensure postdocs reap the career development benefits they should be receiving.

First, for postdocs who have high agency and high goal clarity (the most common pattern among participants in this study), we see the need for in-depth research utilizing a strengths-based approach to more deeply investigate the factors (intrapersonal, interpersonal, and contextual) that have enabled them to become successful agents of their own career development. Since literature suggests that postdocs face a number of challenges related to career progress due to the unique nature of the postdoc role (e.g., Åkerlind, 2005; Camacho & Rhoads, 2015; Chen, McAlpine, & Amundsen, 2015; National Academies, 2014; Puljak & Sharif, 2009; Scaffidi & Berman, 2011), the experiences of this agentic and goal-directed group of postdocs can add much to the current conversation. If we are able to ascertain how they developed their high levels of agency and codify some of the steps they took to position themselves for present and future career coherence, we can develop a set of best practices for institutional postdoc offices—guidelines for institutions as they develop structures to support postdocs, for postdoc offices to provide professional development, and for mentors and supervisors to support their postdocs.

At the same time, we cannot assume that postdocs will have guaranteed success as factors outside of the postdocs' control—notably the highly competitive faculty job market and challenges to the foundations of the traditional academic setting arising from the current political climate—will also shape their career paths, especially for those planning to pursue faculty careers. For example, if the job market shifts for a particular discipline or set of disciplines, educators and administrators must engage all postdocs—even the most agentic individuals—in clearly defining career paths and considering alternative career options (both of which involve enhancing postdocs' career literacy and career context), as well as planning and executing steps needed to attain them (i.e., exercising agency in the form of career gumption).

Postdocs located within the high agency–low goals pattern may be the most critical group to address with supportive interventions, as this group is likely to grow as the academic job markets change. While careers outside of the faculty were once considered "alternative," the dearth of available faculty positions has created a new reality for postdocs, one in which alternative career paths (e.g., to industry or administration) are now commonplace and often a more feasible option for those in STEM fields (Bonetta, 2010; National Academies, 2014; Puljak & Sharif, 2009). As a 2014 report by the National Academies notes,

> *for many the postdoctoral researcher position is becoming the default after the attainment of the Ph.D., especially in the life sciences, and the decision to seek a postdoctoral position is often made without regard to whether advanced training in research is really warranted for the individual in question. (p. 61)*

Therefore, we must reconsider how we provide career development support to postdocs that accounts for the new realities of the labor market facing STEM doctorates if we wish for the postdoc role to remain relevant within academia (National Academies, 2014). Most institutions of higher education have career development centers, but many of these centers primarily serve the needs of undergraduates. As the number of graduate students and postdocs increases, institutional leaders should consider how career development centers can effectively serve these populations that are also in need of career development support. Career centers and their services are often decentralized in large institutions (which are the institutions with the largest populations of postdocs), and thus a particular college (e.g., a college of engineering) may employ their own career development specialists. This model could work for graduate students and postdocs as well, as long as these specialists are prepared and trained to serve populations beyond undergraduates. Institutional postdoc offices should partner with career development centers to share best practices, and career development centers must offer support for individuals who have completed a terminal degree but still need to clearly define their career goals.

Finally, it is important to recognize that for a variety of reasons, postdocs may lack defined career goals as well as the agency needed to move forward on a career path. It is the responsibility of both the postdoc's supervisor/PI and the employing institution to guide the postdoc to success. The postdocs in this study who had high levels of agency often chose strategic career-enhancing options without the support of their PI (sometimes in spite of their PI's directive (or lack of directive) and outside of their expected work hours). Yet because the temporary nature of the role often strips postdocs of permanency and authority (Jaeger and Dinin, 2014; National Academies, 2014), postdocs may feel disempowered to engage in agentic actions to support their careers. For others, the limited career-related training they do receive is a "byproduct of work, rather than a planned career development activity" (National Academies, 2014, p. 65).

Postdocs should not have to choose between fulfilling their current responsibilities and positioning themselves for future career success; rather, because the postdoc role is, by definition, a period of mentored training, career development should be an integral part of the job responsibilities of all postdocs, regardless of field. For example, it is a disservice to postdocs as well as the higher education enterprise as a whole if a postdoc seeking a faculty career is unable to leave his or her lab to gain teaching experience. Although National Institutes of Health and National Science Foundation have instituted individual development plans for postdocs to support their career development, the majority of participants in this study indicated that they and their PIs did not utilize these plans. As discussed in our previous recommendation, supervisors/PIs and institutions both have a responsibility to provide the postdocs they employ with career development support that enables them to define and clarify their professional goals as well as to develop and exercise agency needed to realize those goals, whether through use of a thoughtfully designed individualized development plan or via other avenues.

CONCLUSIONS

Postdoctoral appointments are fundamentally training positions and should enhance career development. However, many postdoctoral scholars report feeling stymied by their position rather than supported (Jaeger et al., 2014). Our study applied concepts drawn from Magnusson and Redekopp's (2011) career coherence model as well as agency theory (O'Meara et al., 2011, 2014) to better understand and explain the career development experiences of STEM postdocs. The majority of postdocs in our study demonstrated a readiness to take agentic action to advance their professional development and work toward their career goals. Yet nearly half lacked clear career goals, including some of those with high levels of agency. Our findings echo those of others who have noted a lack of career path clarity among postdocs (e.g., Bonetta, 2010; Gibbs, McGready, & Griffin, 2015). Our recommendations for practice build on new knowledge about the role of agency and goal clarity of postdoctoral scholars and support intentional career development for this growing population of academics.

ACKNOWLEDGMENTS

The authors extend their gratitude to the initial team of doctoral students including Jingjing Zhang, Ashley Grantham, Allison Mitchall, and Alessandra Dinin lead by their faculty advisor, Audrey Jaeger, one of the authors of this chapter, who gathered the data for this research project. Their role in the postdoc project was essential.

REFERENCES

Åkerlind, G. S. (2005). Postdoctoral researchers: Roles, functions and career prospects. *Higher Education Research & Development*, *24*(1), 21–40.

Archer, M. S. (2003). *Structure, agency and the internal conversation*. Cambridge, UK: Cambridge University Press.

Bonetta, L. (2010). The postdoc experience: Taking a long term view. *Science*, *329*(5995), 1091–1098.

Camacho, S., & Rhoads, R. A. (2015). Breaking the silence: The unionization of postdoctoral workers at the University of California. *The Journal of Higher Education*, *86*(2), 295–325.

Chen, S., McAlpine, L., & Amundsen, C. (2015). Postdoctoral positions as preparation for desired careers: A narrative approach to understanding postdoctoral experience. *Higher Education Research & Development*, *34*(6), 1083–1096.

Creswell, J. W. (2007). *Qualitative inquiry and research design: Choosing among five approaches* (2nd ed.). Thousand Oaks, CA: Sage Publications.

Cresewell, J. W., & Plano Clark, V. L. (2011). *Designing and conducting mixed method research*. Thousand Oaks, CA: Sage Publications.

DeCuir-Gunby, J. T., Marshall, P., & McCulloch, A. (2011). Developing and using a codebook for the analysis of interview data: An example from a professional development research project. *Field Methods*, *23*(2), 136–155.

Gibbs, K. D., McGready, J., & Griffin, K. (2015). Career development among American biomedical postdocs. *Cbe-life Sciences Education*, *14*(4), 44.

Jaeger, A. J., Griffin, K., Huang, B., Cavanaugh, N., Dinin, A., Mitchall, A., et al. (November 2014). The postdoctoral landscape: Collaborating through research and practice to address policy. In *Papers presented at the association for the study of higher education. Washington, DC.*

Knox, S., & Burkard, A. W. (2009). Qualitative research interviews. *Psychotherapy Research, 19,* 566–575.

Magnusson, K., & Redekopp, D. (2011). Coherent career practice. *Journal of Employment Counseling, 48*(4), 176–178.

Marshall, C., & Rossman, G. B. (2011). *Designing qualitative research* (5th ed.). Thousand Oaks, CA: Sage Publications.

McAlpine, L. (2016). Post-PhD non-academic careers: Intentions during and after degree. *International Journal of Research Development, 7*(1), 2–14.

Mertens, D. M. (2010). Transformative mixed methods research. *Qualitative Inquiry, 16*(6), 469–474.

Miller, J. M., & Feldman, M. P. (2015). Isolated in the lab: Examining dissatisfaction with postdoctoral appointments. *Journal of Higher Education, 86*(5), 697–724.

National Academy of Sciences, National Academy of Engineering, & Institute of Medicine of the National Academies. (2014). *The postdoctoral experience revisited.* Washington, DC: The National Academies Press.

National Postdoctoral Association. (n.d.). *What is a postdoc?* Retrieved from http://www.national-postdoc.org/policy-22/what-is-a-postdoc.

National Science Board. (2016). *Science and engineering indicators.* Arlington, VA: National Science Foundation. Retrieved from http://www.nsf.gov/statistics/2016/nsb20161/#/downloads/report.

Nerad, M., & Cerny, J. (1999). Postdoctoral patterns, career advancement, and problems. *Science, 285,* 1533–1535.

Neumann, A., & Pereira, K. B. (2009). Becoming strategic: Recently tenured university professors as agents of scholarly learning. In A. Neumann (Ed.), *Professing to learn: Creating tenured lives and careers in the American research university.* Baltimore, MD: The Johns Hopkins University Press.

Opdenakker, R. (2006). Advantages and disadvantages of four interview techniques in qualitative research. *Forum Qualitative Social Research, 7*(4), 11. Retrieved from http://nbn-resolving.de/urn:nbn:de:0114-fqs0604118.

O'Meara, K., Campbell, C., & Terosky, A. (2011). Living agency in the academy: A conceptual framework for research and action. In *Paper presented at the Annual Conference of the Association for the Study of Higher Education.*

O'Meara, K., Jaeger, A., Eliason, J., Grantham, A., Cowdery, K., Mitchall, A., et al. (2014). By design: How departments influence graduate student agency in career advancement. *International Journal of Doctoral Studies, 9,* 155–179.

Puljak, L., & Sharif, W. D. (2009). Postdocs' perceptions of work environment and career prospects at a U.S. academic institution. *Research Evaluation, 18*(5), 411–415.

Saldaña, J. (2016). *The coding manual for qualitative researchers* (3rd ed.). Los Angeles, CA: Sage.

Scaffidi, A. K., & Berman, J. E. (2011). A positive postdoctoral experience is related to quality supervision and career mentoring, collaborations, networking and a nurturing research environment. *Higher Education, 62*(6), 685–698.

Su, X. (2014). Rank advancement in academia: What are the roles of postdoctoral training? *Journal of Higher Education, 85*(1), 65–90.

van der Weijden, I., Teelken, C., de Boer, M., & Drost, M. (2016). Career satisfaction of postdoctoral researchers in relation to their expectations for the future. *Higher Education, 72*(1), 25–40.

Wei, T. E., Levin, V., & Sabik, L. M. (2012). A referral is worth a thousand ads: Job search methods and scientist outcomes in the market for postdoctoral scholars. *Science and Public Policy, 39*(1), 60–73.

Chapter 7

European Cross-National Mixed-Method Study on Early Career Researcher Experience

Montserrat Castelló[1], Kirsi Pyhältö[2,3], Lynn McAlpine[4,5]
[1]*Universitat Ramon Llull, Barcelona, Spain;* [2]*University of Helsinki, Helsinki, Finland;*
[3]*University of Oulu, Oulu, Finland;* [4]*University of Oxford, Oxford, United Kingdom;* [5]*McGill University, Montreal, QC, Canada*

Chapter Outline

Introduction   144
  The European Context as an
  Influence on Postdoc Career
  Opportunities   145
A European Cross-National
Research Program   146
  Research-Related Experiences
  of Early Career Researchers
  Survey   147
  Multimodal Interview Protocol 149
  Journey Plots   150
  Network Plots   151
Writing as a Primary Form
of Research Communication   152
  Post-PhD Researchers' Writing
  Perceptions   153
  Development of Writing
  Perceptions   154
The Study: Exploring Postdocs
Writing Perceptions   **154**
Method   **154**
  Data Collection   155
  Participants   157
  Analysis   157

  Quantitative Analysis   157
  Qualitative Analysis   158
Results   **158**
  Quantitative Analysis   158
    Postdoc Researchers'
    Writing Profiles   159
    Writing profiles and research
    engagement, experienced
    cynicism and exhaustion,
    supervisory, and researcher
    community support   159
  Qualitative Analysis   159
    Key experiences related
    to written research
    communication   159
Discussion   **165**
  Writing   165
  Strengths and Limitations of
  the Research Approach   166
    First, the Development of
    a Consistent Conceptual
    Framework Shared and
    Assumed by the Different
    Members of the Team   167

The Postdoc Landscape. https://doi.org/10.1016/B978-0-12-813169-5.00007-0

Second, the Creation of Cross-National and Cross-Linguistic Methodological Instruments 167
Third, the Creation of Robust Qualitative Analysis Procedures Across Countries 168
Appendix A **169**
FINS-RIDSS Interview Protocol **169**
Part I: Set-Up and Briefing (Total Time: About 5′) 169

Part II: Initial General Questions (Total Time: About 5–10′) 170
Part III: Journey Plot Instrument (Total Time: About 15–20′) 170
Part IV: Network Plot (Total Time: About 15–20′) 171
Part V: Closure (Total Time: About 5′) 172
References **172**

INTRODUCTION

There is life after earning a PhD and, although research has not devoted great attention to it, this is a complex life, especially for those in a post-PhD research position at a university. Much of this complexity emerges from changes in higher education systems over the past few decades, specifically related to the global economic crisis, which has forced new economies on university and research funding. The impact has been especially felt by recent PhD graduates seeking research-teaching positions. They struggle with the demands of global mobility, interdisciplinarity, and the lack of stable or permanent positions, all of which can lead them ultimately to consider alternate careers (Åkerlind, 2005; Woehrer, 2014). Academe has become a risky career field with uncertain outcomes for those who wish to enter. At an international level, more than half of doctoral graduates leave the academy whether by choice or lack of opportunity, and a number shift their career intentions away from academia as they see the demands placed on their supervisors (McAlpine & Amundsen, 2016). Those who aspire to stay in academia soon learn that postdoc fellowships are at a premium. As a result, many take on salaried research positions working to achieve the goals of principal investigators (PIs) even though such work may not be congruent with their own research interests. Postdocs feel that if they wish to remain in the academy, they are under enormous pressure to develop a unique research trajectory as recognized through publications and sometimes small research grants.

In this chapter, we begin by describing the European context and contrasting it with the North American one. Then, we describe our mixed-method European cross-national research program (Spain, the United Kingdom, Finland, and Switzerland) in which the aim is to better understand the challenges postdocs confront in developing their career trajectories. A number of distinct analyses have emerged from this research program; see, for instance, Chapter 8. Next, we focus on a particular analysis related to how participants with different writing profiles (identified in the survey) deal with common experiences involving written research communication (analyzed through multimodal interviews). We finish by discussing the strengths and limitations of the research approach.

The European Context as an Influence on Postdoc Career Opportunities

The European context is unusual in that there are two levels of governance, both national and Europe-wide. The European goal is to generate consistent patterns and procedures to ease mobility and transferability of individuals, knowledge, and products. Therefore, a key goal in creating the European Higher Education Research Area has been to promote mobility (highly valued for early career researchers (ECRs)), to enhance research collaboration across countries, and to demonstrate globally the strength of European research work. For instance, 7% of the total reported publications by European Research Council grantees are among the 1% mostly highly cited publications in the Scopus database (as of June 2016, Elsevier Scopus database, OpenAIRE database). Another example of pan-European influence is the European Open Science and European Open Science Agenda which all countries are expected to work toward. A further example is the common understanding of researcher development wherein post-PhD researchers are seen as a relatively accomplished group of ECRs. That is, based on the "European Commission European Framework for Research Careers," they are seen as R2 Recognized Researcher—PhD holders not yet fully independent, who are seeking to move to R3 Established Researcher, those with a degree of independence: https://era.gv.at/object/document/1509.

Further examples of the intertwining governance systems are policies as well as research and fellowship funding available from national and European sources to support the development of postdoc career trajectories. For instance, the European Research Council offers funding for basic research for all researchers as well as fellowship funding for ECR through Marie Sklodowska-Curie actions: http://ec.europa.eu/research/mariecurieactions/apply/calls_en. Further, many countries have research councils or trusts that offer a series of research career fellowships. For instance, in the Netherlands, there is a series of grants called *Veni, Vidi, Vici*, focused at different stages of early career development and designed to boost innovative research, promote mobility, and enhance researcher independence and leadership: http://ec.europa.eu/research/mariecurieactions/apply/calls_en.[1]

Despite the variety of fellowship funding sources in Europe, they are, as elsewhere, limited in number. So, as with elsewhere in the world, most who seek to remain in the academy are funded through PI grants on contracts of limited duration. In other words, if we look more deeply into to the work context, a further influence on most postdocs' experience is the nature of the work they are hired to do for the PI. These responsibilities are, in turn, influenced by national research funding policies (McAlpine, 2016).

---

1. The nature of the relationship between these two governance levels is now under discussion given the UK referendum on leaving the EU. It is not yet known what impact this will eventually have, but there are concerns about researcher mobility and access to European research funds.

For instance, research regimes in the sciences and engineering in the United Kingdom and the United States differ substantially. Cantwell (2011) examined the two systems and reported that, in the Unites States, PIs are viewed as "entrepreneurs." When they submit a research proposal, their proposed research plan can be changed as the work progresses, as can their initial plan for hiring researchers to help carry out the work. In other words, such decisions are not limited by research council or institutional policies. In fact, they may be rewarded for a sound training plan for ECR as is the case in the Canadian context. In contrast, in the United Kingdom, PIs are seen as "project managers." The proposed projects have defined goals that must be met. Further, whatever staffing specifications have been set out in the proposal must be maintained. In other words, UK research councils limit how money can be used, which constrains the ability of the PI to act differently once the grant is under way. That is, in North America, they are free to encourage their postdocs to follow up a novel or unexpected finding, whereas in the United Kingdom this would not be feasible.

A EUROPEAN CROSS-NATIONAL RESEARCH PROGRAM

Within the described European context, our cross-national research program started in 2013, with three countries initially involved (Spain, the United Kingdom, and Finland). Our goals were to better understand the challenges researchers confront in developing their career trajectories, a process often characterized as "identity development" (e.g., Kamler & Thomson, 2008). In the last 2 years, other countries such as Switzerland and Denmark have joined the project, and plans for collecting data in France are also under way. The research program spans both PhD students and postdoc[2] researchers and we refer to them collectively as ECRs since they are still developing their research independence (Castelló Kobayashi et al., 2015; Laudel & Gläser, 2008). Including both participant groups in the study, which is rarely undertaken by higher education researchers, enables us not only to compare across national systems but also to compare across the two groups to gain insight into the characteristics of the transition from being a PhD student to a graduate. This is critical if we wish to develop a more complete picture of researcher identity development through their career trajectories, especially given the few theoretically driven studies of this kind. Lastly, by comparing ECR perceptions across national contexts, we can contribute to the literature while also designing European supranational innovative training proposals; additionally, we can develop policy-based recommendations, which constitute two additional aims of the research program.

---

2. We characterize postdocs as researchers who are not yet considered fully independent (despite their skills and expertise), in the second stage of their development as researchers (R2) using the European Commission European Framework for Research Careers. The majority of these postdocs, referred simultaneously as post-PhD researchers or ECRs, tend to be on contract working on others' projects, though a small percentage hold their own fellowships.

Concerning methodology, in our research program, we use a mixed-method design that includes collecting a multimodal set of data combining both cross-sectional and longitudinal approaches (McAlpine, Pyhältö, & Castelló, submitted). In the first phase, the *Research-Related Experiences of ECRs multilingual survey* was developed to better understand ECRs' perceptions of different variables that research has shown are crucial in researcher identity development and career trajectories. The second phase consisted of characterizing participants' understanding of the key or most significant events they experienced over 1 year after participation in the survey. To this purpose, we designed multimodal interviews in which visual tools such as *Journey Plots* and *Network Plots* were integrated. In designing the instruments, our goal was to clarify the constructs we were examining (theoretically and empirically); to generate protocols that drew on previous studies; and to ensure that we explored each construct in different and complementary ways. We wanted to move beyond the use of single methods to be able to contrast as well as integrate different but parallel types of data. Specifically, we perceived the qualitative information as serving not only to explain some variation in the quantitative results but also to generate new questions that might force us to revise previous results and associated instruments. Once these protocols were developed, they were piloted and revised as needed. The main characteristics of the design, development, and use of both instruments—the Research-Related Experiences of ECRs survey and the multimodal interview—are explained in the following text.

Research-Related Experiences of Early Career Researchers Survey

We aimed to design a survey that would explore relevant areas of ECR experiences; one useful for ECRs at different stages; that was not excessively long or time-consuming; and one that was able to allow for different types of responses. To deal with these challenges, we extensively reviewed the research and relied on empirically proven previous instruments (Lonka et al., 2014; Pyhältö, Vekkaila, & Keskinen, 2015; Pyhältö et al., 2016). The process was relatively easy when looking at items from studies of PhD students, but it was much more complicated to find prior studies of post-PhD researchers, where there has been less conceptually based quantitative research. Since we expected both versions of the survey (for PhD and post-PhD researchers) to incorporate the same constructs and be used in combined studies, we decided to modify and adapt the existing PhD students items to suit the researcher context.

The final survey contained seven scales (Likert-type ranging from 1 to 7), along with 17 sociodemographic background questions and 18 multiple choice or open-ended questions regarding specific features of their experience (such as significant events in their trajectory, number of publication, abandonment intentions, and satisfaction).

The two parallel versions (PhD and post-PhD versions) were simultaneously developed in three languages: Spanish, Catalan, Finnish, and English, resulting in six parallel surveys. In this process, three cross-cultural teams consisting of researchers from each of the countries in which the survey was been developed discussed all the items in English, and then adapted them into Spanish and Finnish using forward–backward translation shared and discussed by researchers who were native speakers. Pilot studies were conducted using the three language versions to confirm the appropriateness of the wording–both linguistically and culturally– and the questionnaire structure. At this point, the two versions of the Research-Related Experiences of ECRs questionnaire have also been translated into French and Danish following the same method of forward–backward translation and piloted in Switzerland and Denmark (you can find the English version following this link: https://goo.gl/TlEbpS). Analysis performed to identify the internal structure revealed the six scales have high reliability (alphas were between $\alpha = 0.728$ and $\alpha = 0.929$) and identical factors in the PhD and post-PhD versions (see Castelló, McAlpine, & Pyhältö, 2017; Pyhältö, McAlpine, Peltonen, & Castelló, 2017).

Altogether items included in the seven scales, multiple choice, and open-ended questions addressed the following constructs (McAlpine et al., submitted):

- *Engagement and interest in research*: These constructs accounted for the degree of commitment and dedication to research, as well as interest of ECRs in community, finding and improving their work position or personal development. They were assessed in two separate scales, namely engagement and interest (Pyhältö et al., 2016; Vekkaila, Pyhältö, & Lonka, 2013, 2014).
- *Scholarly communication*: It addresses different ways in which ECRs communicate research findings and was assessed through multiple choice items regarding number and type of publications and conference presentations, open-ended questions regarding significant events in which communication-related events can be described as well as associated emotions and involved people. A scale on writing perceptions was also included (Castelló et al., 2017; Lonka et al., 2014).
- *Social support*: This construct refers to the resources both perceived to be available and used by ECRs in their social environment. It entails formal and informal relationships, both dyadic and group relationships within the researcher community, with peers, supervisor(s), other staff members (Vekkaila, Virtanen, Taina, & Pyhältö, 2016) as well as research groups, international researcher networks, or special interest groups, and relationships with institutional representatives. Social support is a metaconstruct comprising emotional, informational, and instrumental forms (Cobb, 1976; House, 1981; Väisänen, Pietarinen, Pyhältö, Toom, & Soini, 2016). To assess it, we developed two scales entailing *supervisory support*, and *researcher community support* (Pyhältö et al., 2017, 2015).
- *Research conceptions*: In this case, we were interested in how ECRs understand research, the meaning they give to research as well as the final aim they attribute to their own activity as researchers. We developed a specific scale based on results of some previous research mostly on doctoral students (Åkerlind, 2008, 2005; Brew, 2001).

- *Burnout and stress:* They refer to work-related syndromes that develop gradually as a result of extensive and prolonged work-related stress. Burnout has two distinctive aspects: exhaustion and cynicism (Maslach & Leiter, 2008; Maslach, Schaufeli, & Leiter, 2001). Exhaustion is characterized by a lack of emotional energy, and feeling strained and tired at work, while cynicism refers to alienation, distancing oneself from work, losing interest in one's work. Both aspects were assessed through a burnout scale included in the survey (Pyhältö et al., 2016). In addition, stress was measured with one-item stress scale (adapted from Elo, Leppänen, & Jahkola, 2003).

- *Agency and regulation*: These items refer to what extent ECRs perceive freedom to act and to intervene and, in some cases modify what they do, think, and feel as researchers; both are socially situated activities that involve not only personal but also social contexts specially when confronted with challenges (Castelló, Iñesta, & Corcelles, 2013a,b; O'Meara et al., 2014; Reybold & Alamia, 2008). In the survey both constructs were assessed by means of open-ended questions in which participants were asked to explain their most positive and negative experiences, as well as associated emotions and who was involved. When challenges or problems were described they were also prompted to explain how do they act to cope with the situation and, if any, who helped them.

- *Personal life*: The construct was included to assess how ECR experience relationship between research work and personal life conditions (McAlpine, Amundsen, & Turner, 2014). Although we did not develop a specific scale on the role of the personal as an influence on work experience, two open-ended questions were included at the end of the survey.

The first, second, third, and fifth constructs form the basis of the analysis described later in this chapter.

Multimodal Interview Protocol

The multimodal interview was designed to obtain information about the same constructs included in the questionnaire but from a different perspective that emphasizes participants' narrative voices regarding their researcher trajectories. This suggested examining the aforementioned constructs in relation to (1) the significant events participants experienced in the last 12 months, i.e., since they answered the questionnaire; (2) agency and emotions associated with those significant events; and (3) individuals or groups with which participants interact and that shape their network.

After several discussions regarding the rationale, structure, and organization of the multimodal interview, a highly detailed protocol was developed in the three languages (Spanish, English, and Finish; see Appendix A for a reduced version of the English version) fostering consistency in collecting the qualitative data. This protocol describes the four differentiated parts of the designed interview—aims, prompts, and associated questions—as well as when, how, and why visual methods were integrated and contribute to the interview overall aim. Thus, a distinctive feature of this protocol is that it also contained information about the aims of each part and prompts to facilitate the participants' integration of their narrative explanations with the information provided in visual methods as well as in relation to the answers they provided in the questionnaire on the same constructs.

A preliminary step, before conducting the interview, consisted of looking at sociodemographic data provided in the survey. This included age; discipline; career stage; current status; the significant events each participant mentioned in the survey; feelings associated and attempted solutions; results of the scales (social support, interest and engagement, writing conceptions, research conceptions, and burnout); and answers regarding publication experience, international mobility, work expectations, satisfaction, and stress. We completed a participant grid with all this information to provide a basis for asking interviewees to expand previous answers as well as to detect changes, inconsistencies or confirmation between the questionnaire and interview information.

Two specific visual methods were included in the protocol accounting for multimodal data collected. Use of visual methods when accompanied by participants' written or verbal explanations can aid the researcher's analysis by capturing participants' own interpretation of their work (Buckingham, 2009; Miller & Brimicombe, 2003). To enhance the integration of the narrative description of significant events with their temporal distribution and the affective value participants attributed to them, we used an adapted version of *Journey Plots* (McAlpine, 2016). To explore ECR relationships with other individuals, groups, or institutions, as well as to know more about their position in the research community, we designed a new instrument called *Network Plot* (McAlpine et al., submitted; Sala-Bubaré & Castelló, 2017).

Journey Plots

Journey Plots are a visual data collection method particularly useful when the purpose is to understand how significant experiences and related emotions evolve through time (McAlpine, 2016). The journey plot template we used (see Figs. 7.3–7.5, in the results section) showed the progress of time, normally 1 year since this was the period between participants answering the survey and being interviewed. Time by month was recorded on the horizontal axis from left to right and the variation in related emotion from high to low on the vertical axis (top to bottom), with the midpoint marked.

Journey Plots were used in the middle of the interview, approximately 15 min after starting and participants were given the following instructions: "*We will use this tool to represent your doctoral/postdoctoral trajectory through the significant events. I'd like you to draw in this graphic the most significant events during the last year. As you see, the Y axis represents the 'emotional Low/High of the experience' from positive to negative. In the X axis you have the time represented in months."* Interviewers asked participants to explain the graphic while they drew and return to the map to elaborate experiences once it was drawn. They also paraphrased participants' descriptions to summarize the events and, if necessary, they added labels to each of them to ensure they understood the meaning participants attributed to each event. It was also important to comment on the intensity of the events (e.g., whether that event was more positive than a previous one) to guarantee shared meanings of the representations regarding experience intensity.

Network Plots

This instrument was designed to obtain visual information regarding participants' research network in the scientific community—that is, the individuals and the groups they interact with. Network Plots were slightly adapted from a previous instrument used with doctoral students, the community plot (Sala-Bubaré & Castelló, 2017). In both cases, besides exploring participants' research network, they concurrently informed us about how they positioned themselves in it. Participants were asked to draw as many circles as they needed to represent individuals or groups and their relationships and to use one of the circles to represent themselves. Circles were of different sizes reflecting the importance interviewees attributed to a certain group or individual in their research network. They were also asked to write the name of the individual or group on each circle (Fig. 7.1).

Network Plots were placed in the middle of the interview immediately after completing the journey plots and were accompanied by the following instructions: *"We will use this tool to represent your research network at this time. You*

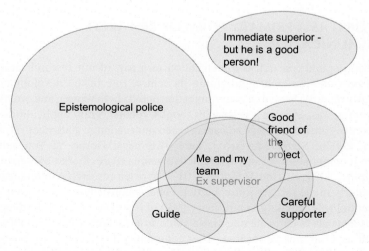

FIGURE 7.1   Example of network plots used in the multimodal interview.

have different circles of different sizes. I'd like you to use as many as you need to represent the individuals and groups that are important in your research experience and the connections between them. The size being the relevance they have in your experience." Interviewers were attentive to asking for relationships between the significant events mentioned in the journey plot as regards the extent to which those individuals or groups were involved in any of these events. After composing their network plot, participants were prompted to think about the extent to which they were satisfied with their current position in the network and to reflect on their ideal or desired position in the near future.

Results from the specific studies within our research program published so far indicate that challenges related to scientific communication were present in most of both positive and negative events ECRs mentioned as highly relevant in their research trajectory (Corcelles, Cano, Liesa, González-Ocampo, & Castelló, submitted). Moreover, recent results show that although almost all postdocs have experience in scientific publication, some of them still hold maladaptive conceptions regarding what writing is and how to deal with it (Castelló et al., 2017). This appeared to be related to a lack of engagement and lack of perceived support from the community. It may also be a result of high levels of burnout clearly affecting their career trajectories and ultimately the development of their researcher identity (McAlpine et al., submitted). Based on the relevance and novelty of these results, in the next sections we first claim the relevance of writing for postdocs as a primary form of research communication, then review what we already know regarding writing perceptions and their development and finally explore the existence of postdocs' writing profiles (identified in the survey) and how these profiles are related to how the participants deal with common experiences involving scientific communication (analyzed through multimodal interviews).

WRITING AS A PRIMARY FORM OF RESEARCH COMMUNICATION

We incorporated research communication as a part of our research program given previous research highlighting its central role in the development of scholarly identity. Research communication entails both written and oral forms of reporting research such as articles, books, reports, grant applications, and conferences presentations, for academic and nonacademic audiences (Hyland, 2002a, 2002b; Starke-Meyerring, Paré, Artemeva, Horne, & Yousoubova, 2011). It comprises a variety of distinctive genres (Paré, Starke-Meyerring, & McAlpine, 2009; Russell & Cortes, 2012). Written research communication, particularly in the form of journal articles and scientific books, plays a unique and central role, not only as a primary form of reporting research findings but also as a means to cultivate one's own voice as a researcher while establishing oneself as a full member of researcher community. Particularly, for post-PhD researchers who are still moving toward independence, writing is a central site of identity construction as each individual is positioning the self as a legitimate voice with a contribution to make (Cameron, Nairn, & Higgins, 2009; Castelló et al., 2017; Kamler & Thomson, 2008; Lea & Stierer, 2011). In addition, publishing, especially in highly ranked peer-reviewed journals, has become a widely accepted measure of research productivity that strongly determines an ECR's career trajectory within the academy. Research productivity has been shown to contribute to postdoc career development (Kyvik, 2013; Ross, Greco-Sanders, Laudenslager, & Reite, 2009). More recently, research has shown that postdocs have difficulties in managing writing scientific articles (Ross, Greco-Sanders, & Laudenslager, 2016) both in their own language or when using English as a second language (Bazerman, Keranen, & Encinas, 2012). It seems that only a minority of recent postdocs had extensive experience in publishing in international journals (Castelló et al., 2017). Surprisingly, research on individual variation between the post-PhD researchers in terms of their writing perceptions is still scarce, and even less is known about the key experiences contributing to the development of such perceptions among post-PhD researchers.

Post-PhD Researchers' Writing Perceptions

Post-PhD researchers' writing perceptions comprise their mental representations concerning scientific writing and the practices and habits they develop around writing activities (Castelló et al., 2017; and in the case of doctoral students, Lonka et al., 2014; Torrance, Thomas, & Robinson, 1994). Post-PhD researchers' writing perceptions can be grouped based on their contribution to effective text production into *adaptive* comprising of perceptions (and resulting actions) that are functional in advancing researcher's writing goals, and *maladaptive*, referring to less functional perceptions,which limit the potential for individuals to act in ways that advance their writing goals. *Adaptive perceptions* consist of *productivity* (Pyhältö, Nummenmaa, Soini, Stubb, & Lonka, 2012) and perceiving writing as a way to *create knowledge* (Bereiter & Scardamalia, 1987). There is evidence that *adaptive writing perceptions* are related to productivity in terms of number of publications (Castelló et al., 2017). *Maladaptive perceptions of writing*[3], on the other hand, include *blocks* (Rose, 1980), *procrastination* (Lonka et al., 2014), *perfectionism* (Boice, 1993), and the perception of writing as an *innate ability* (Sawyer, 2009), the result being that researchers may avoid investing in functional activities that would advance their writing.

---

3. We are mindful of the pejorative nature of the word, maladaptive, despite it being a relatively well known and studied psychological construct. Our interest in it is to understand how to help researchers recognize less functional strategies and shift their perceptions to lead to more functional behaviors.

Development of Writing Perceptions

Post-PhD researchers' perceptions of writing do not develop in a vacuum; rather, they are highly affected by the quantity and quality of research community interactions regarding written research communication. For instance, supervisory and researcher community support or lack of it can either promote learning of adaptive writing perceptions or hinder it. Post-PhD researchers' writing perceptions were positively related to support from their supervisor and researcher community and negatively to their risk for developing burnout (Castelló et al., 2017). Also, disciplinary writing practices of the community such as coauthoring regulate post-PhD researchers' opportunities to learn from more advanced colleagues. The nature and quality of feedback on one's texts is likely to affect post-PhD researchers' willingness to seek feedback in the future (Florence & Yore, 2004).

THE STUDY: EXPLORING POSTDOCS WRITING PERCEPTIONS

This study aimed to gain a better understanding of individual variation in post-doc researchers' perceptions of academic writing and development of such perceptions. The following research questions were addressed:

RQ1: What kinds of writing profiles can be detected among post-PhD researchers?

RQ2: Do the writing profiles differ with regard to (1) supervisory and researcher community support, (2) experienced burnout, (3) research engagement, (4) productivity, and (5) abandonment intentions?

RQ3: What kinds of key experiences related to written research communication did post-PhD researchers report? To what extent were these experiences representative of their writing perceptions and writing profiles?

METHOD

Data were collected in late 2015 and early 2016 and a convergent mixed-method design was applied (Creswell & Plano-Clark, 2011). That means that, in this case, quantitative and qualitative data were analyzed separately and were then combined at the intermediate stage of the study. Quantitative data were utilized to better understand the variation in post-PhD researcher perceptions. Qualitative data were used to explore, confirm, and expand on the quantitative findings by analyzing the significant experiences related to written research communication that postdoc researchers' reported. By combining these different sets of data, it was possible to gain a more comprehensive understanding of writing as a form of research communication (Bryman, 2006; Creswell & Plano-Clark, 2011; Greene, Caracelli, & Graham, 1989).

Data Collection

Research-Related Experiences of ECR survey data were collected by e-mail through an online survey in winter 2015–2016. In this study, we utilized data from the scales on

- *Academic writing* scale, comprising *productivity* (four items), *procrastination* (four items), *blocks* (four items), *perfectionism* (four items), *knowledge creation* (three items), and *innate ability* (three items) (Castelló et al., 2017; Lonka et al., 2014)
- *Supervisory and researcher community support scale* (nine items) entailing *supervisory support* (four items) and *researcher community support* (five items) (adapted from Pyhältö et al., 2015, 2016)
- *Abandonment intentions* (one item), *experienced burnout* (total eight items), including *cynicism* (four items) and *exhaustion* (four items), *research engagement* (five items) (see scales in Table 7.1 and more detailed information in Castelló et al., 2017).

TABLE 7.1 Descriptive Statistics on Writing, Supervisory and Researcher Community Support, Burnout, Engagement, and Publication Experience

Items/Scales	N of Items	Alpha	M	SD	Min	Max
Knowledge creation	3	0.76	5.63	1.22	1	7
Productivity	4	0.86	3.79	1.41	1	7
Procrastination	4	0.77	4.06	1.42	1	7
Perfectionism	4	0.78	3.05	1.37	1	7
Innate ability	3	0.72	2.05	1.11	1	7
Blocks	4	0.78	3.07	1.28	1	7
Supervisory support	4	0.95	5.07	1.81	1	7
Researcher community support	5	0.88	5.11	1.25	1	7
Cynicism	4	0.87	2.92	1.58	1	7
Exhaustion	4	0.88	3.54	1.42	1	7
Engagement	5	0.93	5.43	1.29	1	7

TABLE 7.1 Descriptive Statistics on Writing, Supervisory and Researcher Community Support, Burnout, Engagement, and Publication Experience—cont'd

Publication Experience	Journals		Total (%)
	First Author	Coauthor	
No experience	6	13.5	27.6
1–2 publications	28.5	31.1	26.5
3–4 publications	17	20.7	20.4
5–6 publications	19.5	10.4	10.7
More than 7	29	24.3	14.8

In addition, *publishing experience* and *abandonment* intentions were documented:

- *Have you already published research? If yes, how many publications you have in peer-reviewed journals: as a first author/not as a first author?*
- *Have you considered dropping out of your researcher work?* If yes, please explain briefly why you have considered dropping out of your researcher work.

All participants received written information about the project and gave their consent to participate according to the ethics clearance procedures in the respective jurisdictions.[4] It took 15–20 min to complete the survey and, near the end, respondents were invited to participate in an interview. Where data were missing for key variables, post-PhD researchers were excluded from the analysis.

As mentioned, the interview protocol drew on the same constructs as the survey and was structured to elicit more in-depth information about participants' experiences surrounding key concepts. Three points are important for this study: first, we sought to learn more about their broader lives; second, we asked them to expand on their responses to certain items in the survey; third, we analyzed data from one of the two visual methods, the *Journey Plot*. As mentioned, the *Journey Plot* represented the previous 12 months on the horizontal axis from left to right and the variation in related emotions from high to low on the vertical axis, with the midpoint marked.

Participants were asked to map the key events within their journeys and to reconstruct the journey orally, expanding on how they experienced and

---

4. The project was approved by the Ethics Committee of the University of name omitted for the blind review (CER-URL-2013_005) and by the Spanish Ministry of Economy and Competitiveness (CSO2013-41,108-R).

responded to the emotional highs and lows. In this way, we could see how their emotional resilience, coping strategies, and commitment changed related to the critical incidents they identified. In this study, we analyzed key events related to written research communication.

Participants

Altogether 282 postdoc researchers from research-intensive universities the United Kingdom (n = 98) and Spain (n = 184) responded to the *Research-Related Experiences of ECR* survey (see more detail about sampling strategy in Castelló et al., 2017). Participants from Finland were not included in this chapter because the whole process of data collection—including interviews— was not finished yet in this country when we started to analyze data. Social scientists represented 2/3 (n = 195) and scientists 1/3 (n = 87). Their mean age was 35.9 and just over half (53.0%) were female. The mean time for completing the doctoral degree was 5.2 years; 81% had completed their doctoral thesis in the form of a monograph and 19% as article compilation; the mean time since graduation was 3.5 years, with the majority of participants being less than 5 years from earning the doctoral degree (56.4%). In addition, they were typically either salaried researchers (47%) or competitive grant holders (36.5%). Only 5% were employed outside the university, and 11.5% were unemployed at the time of the data collection.

Qualitative subsample: Among those ECRs who declared their interest in participating in the second qualitative phase of the project, a total of 57 postdocs—11 from the United Kingdom, 16 from Finland, and 30 from Spain— were interviewed. 32 were female and 25 male, with minimum and maximum age at graduation 27 and 57 years (median 30 years). The mean graduation time was 4.9 years.

For this chapter, we chose to examine the experiences of three postdocs who were social scientists in the United Kingdom so they would at least share some common characteristics.

Analysis

Quantitative Analysis

Descriptive statistical analyses were carried out using IBM SPSS Statistics 21. To determine the postdoc researchers' perception of academic writing, experienced supervisory and researcher community support, research engagement, and burnout, means and standard deviations were calculated for each scale.[5]

---

5. The writing scale (Castelló et al., 2017; Lonka et al., 2014), supervisory, and researcher community scale (Pyhältö et al., 2016; Pyhältö et al., 2017), burnout, and engagement scales have been confirmed in previous studies, including both explorative and confirmatory factor analyses to determine the structure of the scales.

To divide the sample into meaningful subgroups according to the writing perceptions, K-means cluster analysis was carried out. In the K-means cluster procedure, the number of clusters is chosen by the researcher, and cases are grouped into the cluster with the closest center. Cluster solutions with two and three clusters were calculated; however, the three-cluster solution gave the most homogeneous profiles and was in line with the theoretical presumptions. One-way-analysis of variance was carried out to explore interrelation between the profiles and experienced burnout, engagement in postdoctoral research, and supervisory and researcher community support.

Qualitative Analysis

All data for each participant, treated as a case, were imported into MAXQDA data analysis software. Parental codes were defined by a team of four researchers to ensure consistency around the overall themes. Each of these parent codes was then reviewed and subcodes created. For this study, we looked at (1) participants' survey responses related to writing perceptions and number of publications, (2) the bios we had created to summarize their lives, and (3) the interview segments in which the significant events entailed written research communication. We analyzed these data in an integrated fashion to develop cameos of how writing played a role in their lives. We have chosen to provide only three cameos here, each one representing the profile that emerged from the quantitative analysis. Each cameo or narrative embeds the individual's significant writing experiences within their work and broader lives.

RESULTS

Quantitative Analysis

On average, postdoc researchers reported high levels of *knowledge creation* and *productivity*, although variation in perceived productivity occurred. At the same time, they rarely perceived writing as an *innate ability*, although they reported suffering moderate levels of *blocks*, *perfectionism*, and acknowledged that they experienced high levels of *procrastination* (see Table 7.1). Postdoc researchers reported receiving high levels of researcher community and supervisory support combined with high research engagement, and low levels of cynicism. At the same time, they reported suffering average levels of exhaustion.

The majority of the researchers (95%) had some kind of publication experience. However, their publishing experience was somewhat limited: 6% had never published as first authors in peer-reviewed journals. Moreover, despite their career phase, one-quarter had hardly any experience, with only one or two publications (which might not include first-author status); one-fifth had three to four publications; and one-quarter had more than five publications both as a first author and coauthor in peer-reviewed journals.

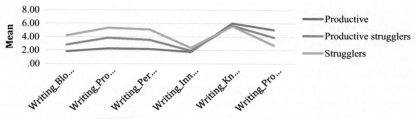

FIGURE 7.2   Postdoc researchers' writing profiles.

Postdoc Researchers' Writing Profiles

Three distinctive writing profiles were identified. The first cluster culled from our analysis was *Productive struggler*. It was the most common profile among the post-PhD researcher (43.4%; *n*=82). (see Fig. 7.2). *Productive struggler*–profile holders reported high levels of productivity, but at the same time, they suffered from medium levels of perfectionism, procrastination, and blocks. The second profile was *Struggler*, representing 30.1% (*n*=57) of the postdoc researchers in the sample. The *Struggler*-profile holders showed high levels of blocks, perfectionism, and procrastination combined with reduced levels of productivity.

The third cluster culled from our analysis was *Productive*. It was the least common profile among the postdoc researchers, representing 26.5% (*n*=50) of our sample, but it was still almost as common as those who fit the *Struggler* profile. Postdoc researchers displaying the *Productive*-profile reported high levels of productivity, and unlike their counter partners with *Struggler* profiles, they did not suffer from blocks, procrastination or perfectionism. The three profiles shared the perception of writing as an ability that is not innate and leads to knowledge creation.

Writing profiles and research engagement, experienced cynicism and exhaustion, supervisory, and researcher community support

Postdoc researchers within the *Productive* profile experienced less cynicism (F(2, 182)=3.232, *P*=.042) were more engaged in their research (F(2, 186)=4.119, *P*=.018) than postdoc researchers within the *Struggler* profiles. However, no statistically significant differences between the profiles were detected in supervisory and researcher community support or in experienced exhaustion.

Qualitative Analysis

Key experiences related to written research communication

The qualitative analysis provided a more nuanced understanding of how researchers' perceptions of writing were related to their work as researchers, as will be evident from the profiles below. We begin with Kelsey, a productive writer; then Geri, a struggler; and finally, Fred, a productive struggler. What emerges from across these cameos representing the three profiles is how much we need to situate post-PhD researchers' writing perceptions, productivity, and significant writing experiences within their broader work lives. Fred's case is particularly interesting since it combines aspects of both producer and struggler. He demonstrates how productive strategies can be used to help cope with some disabling perceptions.

We view Kelsey, 36, as a *productive* writer. She is a post-PhD researcher in the United Kingdom in the social sciences and is unusual in having fellowships rather than being salaried on someone's grant. Since graduating in 2011, she first held a writing fellowship that enabled her to publish a good number of articles and book chapters. After that, she submitted several applications for fellowships and earned yet another one. Kelsey notes that she has a very good fellowship mentor who discusses her research with her. However, she is working in a department where most of the people have research interests that are not related to her research focus, so she mainly works on her own. Instead of within her department, she gets feedback on her work at conferences and from reviews. She views herself on an academic track and is actively managing her career (submitting funding applications (reporting in 1 year 10), applying for jobs, and networking).

With regard to writing, her survey responses characterize her as a regular and productive writer who writes regularly, regardless of her mood; thus she produces a large number of finished texts. She saw writing as a creative activity, a way of creating new ideas and expressing herself, as well as a way to develop her thinking. She stands out for her number of publications: 11 as first author in a peer-reviewed journal or book and 5 not as first author in a peer-reviewed journal or book.

Interestingly, all the significant events (both positive and negative) in Kelsey's journey plot were linked to academic communication (see Fig. 7.3). The significant experiences we explore here relate to three different written genres: grant applications, her book being nominated for an award, and coediting a book. Kelsey described having submitted two grant applications in the previous April and "was happy about that" (positive experience). As she said

> *I like the process of writing funding applications ... the actual proposal and speaking with other people about the proposal and getting a group of people involved ...and thinking about the possibilities.*

Unfortunately, in July, she heard she had not been successful in one, which was a "low" (negative experience). But in August, she was awarded the second submission and "has been living off that" (positive experience). Regarding nomination of the book for an award (positive experience):

> *My book was submitted to [UK scholarly organization] ...nominated for ...a very prestigious prize so I was really pleased. So, I found out in April that I didn't win ...but anyway I was very happy to be shortlisted.*

The last experience we describe here is her work in getting a coedited book to production. We explore her experience in more detail since it is an insightful example of the challenges of a collaborative writing project:

> *It's difficult working with two people who …didn't pull their weight …the strategy I used was I wrote to them and would say things like "Let's divide the tasks or divide the chapters – I'll edit these chapters, you edit these chapters, and then we'll meet and we'll discuss" …that was …the tactic that I most continuously used …I gave them lots of timelines. But …it still didn't work because… For example, we'd meet and then they still hadn't edited their chapters, or if they did, they had done what I would perceive as very poor editing …So, I ended up doing a lot [of further revision].*

> *At the time, I felt very, very, very frustrated …[one, a man] would push himself forward. He…insisted on being the first author on our introductory chapter …but he did such a bad job [drafting] that I ended up re-drafting it. So, that was what irritated me more than anything else. He…they got a lot of [attention] …but… they didn't deserve it [laughing]! … I would never work with those people again …one of the main challenges …that I have experienced in my work, it's difficult to find a good collaborator …I've learnt my lesson and I won't work with them again. But …it's difficult to gauge until you work with someone what it will be like working with that person.*

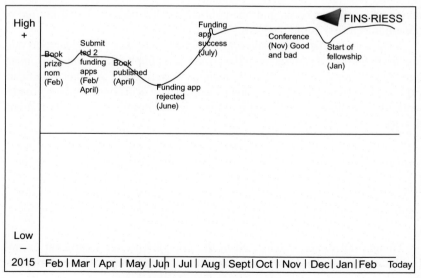

FIGURE 7.3   Kelsey's journey plot.

Different from Kelsey is Geri, 44, who is a *struggler*. She is a social science postdoc who came to the United Kingdom from elsewhere in the European Higher Education Area (English as other language) to complete her PhD. She graduated in 2015 and got a 6-month contract in her PhD department that did not start immediately, so she took a nonresearch job in the university to tide her over. Initially, things went well, but over time, she came to feel it would be impossible to deliver on the expected outcomes. She thought of quitting, but her manager convinced her that she was making good progress and she continued. But now this position has ended. She has been offered another contract for a month and, at the same time, has been applying for jobs, so far without success. She feels quite anxious given the uncertainty of the job situation: "Even though I invest so much …I work on my networks, I publish, I engage in very high level research, but still, it just doesn't come up to a permanent job." As a result, she is considering seeking work outside of academia.

Regarding writing, her survey responses characterize her as someone who struggles to write. Her previous writing experiences have been mostly negative. Perhaps as a result, she often postpones writing tasks until the last moment since she finds it difficult to start writing and will only start if it is absolutely necessary. In other words, without deadlines, she would not produce anything. She also sometimes gets completely stuck if she is required to produce texts and sees herself as too self-critical. Subsequently, she finds it difficult to hand over her texts because they never seem complete, and she could revise endlessly. She has no publications as a first author in a peer-reviewed journal or book, two publications as a first author in a nonpeer-reviewed journal or book, and one publication in a peer-reviewed journal or book not as a first author (see Fig. 7.4).

A significant relatively positive writing experience was writing a report during her first postdoc (though note the effort that is implied): "I *actually …managed* to produce a report …for the [country] Government." Another significant experience related to *not* writing and is connected to decisions about her broader work life. She decided not to apply for fellowship, despite the security it might provide:

> It's very difficult for me to work alone, and this is …the reason why I haven't applied for fellowships …because this is kind of like doing a PhD all over again …you …work on your own …you're really alone …you don't even have a supervisor anymore. And I realized that this is really difficult for me to work …on my own and not having … someone to work with, and so my strategy is …to look for … research projects that are part of a team.

Though not directly connected in the interview to her significant events, Geri's description of her stance around writing aptly summarizes her experiences:

> I enjoy more writing …policy reports or briefing papers than for peer-reviewed journals …for a year now, I haven't written for …a peer-reviewed journal because

my latest submission was rejected and I still don't know what will I do …It's very difficult, to get a rejection, and I know …this is just part of being an academic, to get all these kind of rejections …and I know that the work could be improved …but I'm so afraid of being rejected again …so I have this barrier of sending for publication in peer-review journals …So, basically, writing is not very easy, but …I know what it takes to write – you just have to sit in front of your computer and start writing. It's not going to happen any other way. So, I know how to do that, but I have emotional barriers …in terms of …writing for publication in a peer-reviewed journal.

Fred, 33, is a *productive struggler.* In other words, he shares many of the same views on writing as Geri, but at the same time, he is productive. Like Geri, he is a social scientist who came from elsewhere in Europe (English other language) to complete his PhD, graduated in 2013 and returned to his home country to work as a postdoc. Some months ago, he had two jobs (one related to research, one administrative), but decided to quit the latter because he did not have time and energy for both, especially while achieving a writing goal (see below). He regards being an academic as a vocation. Yet, for him, one of the most difficult things about being a researcher is writing, regardless of which language he uses.

With regard to his perceptions of writing (survey), while he views it as a creative activity that develops thinking and provides a means of new creating ideas

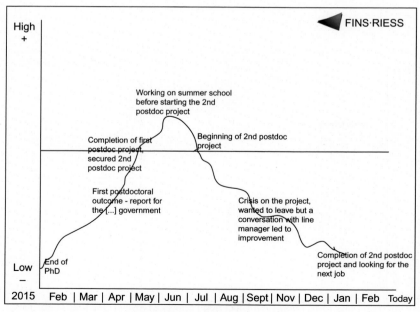

FIGURE 7.4   Geri's journey plot.

and expressing himself, he is somewhat conflicted. For instance, he believes he can generally write regularly regardless of the mood he is in, but he also is too critical, could revise endlessly, and finds it difficult to hand over his texts because they never seem complete. Further, he can produce a many finished texts but sometimes gets completely stuck if he has to produce them. In addition, he not only sees himself generally as a regular and productive writer but also finds it difficult to start writing. Finally, he finds it easier to express himself in other ways than writing. Yet he has 10 publications as a first author in a peer-reviewed journal or book. It seems his enabling strategies help him overcome his disabling perceptions. He is interested and seeks out information about how others write. Also, since he is generally critical of his own work, he works to convince himself that the writing is good enough. As well, he forces himself to write extended notes of what he researches to later connect them and form whole pieces of text.

Perhaps as a result of these strategies, his significant experiences related to writing were all positive (see Fig. 7.5):

[The first] one was when "I submitted a book manuscript and it was accepted … Then the next peak is … when I learned that I was successful in a grant application. [And then he submitted a proposal] and got funding for a big 3-day conference, and it …was very good …successful … What else do you want? A book and big grant success and massive conference…Yeah, it was the best year whatsoever, definitely, in terms of professional success and recognition.

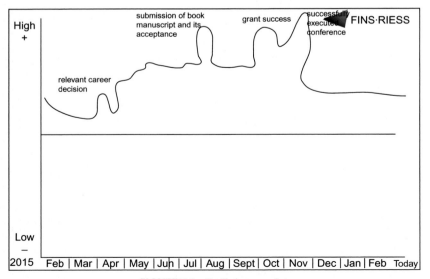

FIGURE 7.5 Fred's journey plot.

Nevertheless, he had had to make a tough career decision in moving forward on the book proposal:

> *The primary thing is that I had to finish the manuscript of the book, and I couldn't work on it [he was working two jobs], so I had to make a tough decision: have a book or …carry on, without ever finishing the book, and keep that sort of additional job, which was very interesting and very cool and I liked it a lot [had many friends and a mentor], but I just simply couldn't…. I didn't have the capacity to combine these two, and the only…the only kind of possibility was to decide either you go, you know, path A or path B – there is no middle ground. You know, there is not a third way. There was no C or anything of that kind. So…so I decided as I decided. I don't regret it …but I miss it sometimes. I miss the person …although we are friends and all that, but it's not like as it was before …And he was a nice kind of guide as well for me.*

DISCUSSION

In this chapter, after describing the European context and contrasting it with the North American one, we first described our mixed-method and cross-national research program in which the aim is to better understand the challenges post-docs confront in developing their career trajectories. Second, we focused on individual variations in postdoc researchers' perceptions of academic writing and their development of such perceptions.

Writing

Among the findings, it is worth mentioning the three writing profiles identified based on the participants' writing perceptions. This is a new contribution to research literature since it differs from the results of previous studies with both doctoral students and post-PhD researchers (Castelló et al., 2017; Lonka et al., 2014) where in the variation in participants' writing perceptions was explained only by two profiles.

The identification of the three distinctive profiles, which we have labeled as *productive*, *productive struggler*, and *struggler*, provides a better explanation for postdocs' understanding of what writing is and how they deal with it than did the two profiles. We are firmly convinced this more nuanced explanation was facilitated by our choice of methods: the convergent analysis of data from the survey and interviews provided us complementary information about the participants' writing perceptions and led us to perform a deeper analysis before reaching a decision regarding writer profiles. Data from the interviews clearly support the existence of more than two profiles while adding interesting, specific information regarding the emotional value of trajectories. As seen in the journey plots of both Kelsey and Fred—the representatives of the *productive* and the *productive struggler* profiles—the significant events they mentioned were all above the midpoint line, whereas in the case of Geri, who was defined as a *struggler*, almost all the events were situated below the line. Fred's case tells us that struggling with writing is not necessarily or always related to negative emotions and lack of written products. Instead, as reported, positive experiences resulting from his intentional use of productive strategies related to publishing were, in turn, perceived as successes in a postdoc trajectory. These positive experiences might counteract the perception of difficulties in managing some aspects of scientific writing process, such as finishing texts (perfectionism), starting to write (blocks), or postponing writing endlessly (procrastination). Further, they would support the continued use of the enabling strategies.

It is interesting to note that our results indicate that some of these variables, such as procrastination and blocks that have been traditionally associated with maladaptive writers (Lonka et al., 2014) can also be related with productivity in some cases. In fact, this is what most of the postdoc in our sample reported in the survey: being productive strugglers.

It is also surprising that we did not find any relationship between these writing perception profiles and perceived support from the community that was previously detected when variable based approach was utilized in exploring the interrelation (Castelló et al., 2017). This is still an open question that would benefit from detailed qualitative and longitudinal analysis to better understand the complex relationship between the social support ECRs perceive from the scientific community, engagement in research and their writing perceptions. This is especially important the case at a time in which coauthorship practices are not only more common but also are requested as a way to display international collaboration and networking—crucial capabilities for ECRs career development.

We are still analyzing qualitative information from the multimodal interviews from the rest of the participants to understand better the relationship between networks and writing, both in terms of processes and products. We reported here information regarding what kinds of key experiences related to written research communication postdocs reported, and the relationship with their writing profiles and associated emotions. We did not look at the network plots to find out more about how these key experiences, writing profiles, and emotions are linked to other individuals or groups that have been relevant in postdocs' research trajectories. Almost all of our participants reported writing and publishing with others in a variety of situations; thus, we consider this type of writing experience deserves a careful analysis which research on ECRs has not systematically performed yet.

Strengths and Limitations of the Research Approach

At the beginning of this chapter, we promised an analysis of strengths and limitations of the research approach adopted in our European research program. Regarding strengths, we would like to highlight that, to our knowledge, this is the first time that a specific research program has been designed with the ultimate goal of creating cumulative evidence of how ECRs develop their careers and acquire their researcher identity across countries, roles, and at different stages of this development. Further, the research program is an example of collaborative research within Europe, which entails a permanent effort in dealing with several issues researchers must consider when doing cross-national, cross-cultural, and cross-linguistic research, as well as cross-paradigms and science traditions. Also, within this context, our respective strengths as researchers have given us the potential to use a mixed-method and multimodal approach.

We now consider the major strengths of the research approach adopted relate to how we addressed the following three challenges:

First, the Development of a Consistent Conceptual Framework Shared and Assumed by the Different Members of the Team

We work in a very big research team that results from the combination of already existing previous national teams that have their own histories, expertise and conceptual frames. We invested a great deal of effort in sharing, discussing, and integrating these conceptual frames in a consistent proposal in both quantitative and qualitative parts of the project. Integration of conceptual frames requires a deep knowledge of literature combined with good analytic abilities and thus benefits from having different types of expertise—both methodologically and conceptually—within the team. At the same time, such collaboration requires planning enough face-to-face meetings as well as producing memos and documents to bring closer different positions and built a shared understanding of key constructs involved in the research program. This process takes time and goes along with building trust to work together at different research stages.

Second, the Creation of Cross-National and Cross-Linguistic Methodological Instruments

As discussed, we developed a parallel survey for doctoral students and postdocs adapted to cultural characteristics of four different European countries and in four languages. Moreover, we created a multimodal interview in which two visual methods were integrated, again in four languages, to be used in five European countries. These were major challenges that needed to be carefully addressed because validity and reliability of data collection and analysis depend on the quality of the instruments designed. In both cases—survey and interview—we relied on some existing national instruments and modified them in several ways to meet the conceptual and contextual requirements of the European research program.

For the survey, this implied starting from some Finish instruments, mostly addressed to doctoral students (Pyhältö et al., 2015, 2016), and moving toward their adaptation to postdoc researchers and ECRs. It also involved the creation of a new scale (research conceptions) as well as the new open-ended section on significant events and several issues related to agency, regulation, and emotions.

For the interview, we started from previous protocols that included *Journey Plots* used with Finnish doctoral students and ECR in the United Kingdom (McAlpine, 2016; Vekkaila et al., 2013) as well as from a protocol including a

Network Plot used with doctoral students in Spain (Sala-Bubaré & Castelló, 2017). These antecedents led us to design what we have called a multimodal interview in which both visual methods are integrated. Having a shared conceptual framework (first challenge) helped us to do this integration. Further, careful piloting in each country was great opportunities not only to refine the instruments but also to deepen our understanding of the key constructs we are trying to assess. And the story goes on, based on first results in which quantitative and qualitative data were brought together (McAlpine et al., submitted), we have already reformulated the research conceptions scale and added more items on personal life in the survey. A new process of data collection has already started in South Africa, Estonia, and Finland that should offer new insights into the usefulness of such changes.

Third, the Creation of Robust Qualitative Analysis Procedures Across Countries

Practices regarding qualitative data analysis often remain occluded. It is not common to detail all the steps from data preparation and organization before the analysis to reliability measures and transparency of the codification process. We addressed this challenge by developing meetings in which procedures were discussed, resulting in detailed protocols for all phases of the analysis, and we shared sessions of practice in coding and solving emerging problems. In this case, the shared used of English as the *lingua franca* helped us to develop common protocols and joint practices.

As for limitations, we are aware that such an approach is highly time-consuming, which can run against the exigencies of obtaining—and publishing—results from research as quickly as possible. However, we hope the accurateness and quality of the reflective process underlying the theoretical and methodological decisions might help us to move forward and make solid contributions to the advancement of research in the field.

Another limitation involves practical and logistic issues. Combining transnational meetings with national teams' development and multiple demands is not easy at all. We need to harmonize institutional and administrative different priorities and agendas to guarantee funding and opportunities for all the researchers included in the research program. Not to be overlooked is their obvious and already mentioned differences in terms of language, nationality, and background including different stages of their careers.

Finally, we acknowledge that despite the initiative being pioneering in collecting and analyzing parallel data at the European level, the scope of the project is still limited. While in some European countries there is an interesting research on ECR experiences that contributes to identity and career development, there are still other countries where this research is practically nonexistent. As noted above, we have already started to broaden the scope of the project to other countries. The next step is to develop the first pan-European proposal to apply to the Erasmus + program devoted to strengthen partnership, joint research, and innovative training across European countries.

Specifically, the ultimate goal of the research program was to develop more integrated theoretical models and training proposals that take into consideration European specificities at the same time that commonalities and challenges are addressed. In line with this, we have already launched some initiatives such as the European Association for Research on Learning and Instruction Special Interest Group on Researcher Education and Careers (EARLI SIG-REaC; see: http://recerca.blanquerna.edu/education-careers-sig/), which has been active since 2014. The SIG Researcher Education and Careers was born in response to the need for a European forum for researchers studying researcher development and aims to promote, encourage, recognize, and support innovative research-based practices that better prepare future researchers for their careers and thus address priorities in graduate and postgraduate education and training. In this regard, the team has started to develop training courses and multimodal materials (e.g., interactive tools and longitudinal questionnaires and videos on critical incidents; see: https://www.fins-ridss.com/) and has applied for European funding to develop an interactive platform in which different training resources for ECRs will be integrated.

In the near future, we see ourselves broadening our research to explore more diverse researcher situations and explore nonacademic research careers in such a way that we can expand the picture of researcher identity development to take into account different types of transitions. Methodologically, we seek to further explore mixed-method approaches in longitudinal designs. This implies looking for alternative data analysis methods that ensure dynamic, valid, and rigorous systems to integrate different types of data evolving through time.

APPENDIX A

FINS-RIDSS INTERVIEW PROTOCOL
Part I: Set-Up and Briefing (Total Time: About 5′)

Activity
Introducing oneself and reminding the interviewee the purpose of the interview
Ask interviewee not to mention third parties by name (e.g., supervisors, colleagues…)
Seek permission to audio and video record
Tell the interviewee that interview transcript will be sent to them and they should add/delete/clarify as they see fit
Switch on the recorders
Reminding interviewee the confidentiality of the data collected and the possibility to finish the interview at any moment.
Collecting the consent form
Ask if any questions before start

Part II: Initial General Questions (Total Time: About 5–10′)

Question/Theme/Prompt	Notes
1a. How is your situation, if at all, different from when you did the survey? 1b. What is your current situation?	We look for changes from the questionnaire answer: funding, research development…
2. What does it mean in terms of your current activities?	
3. Do you remember why do you decided to start a PhD (or to develop a research career)?	Just to start conversation, one general reason can be considered enough
4. What does research means to you? After the participant answer: You considered research as (answer from the questionnaire to the research conceptions items)? (If several options were marked as high, more than 3, then ask: Would you prioritize and explain your answer?) Has it changed? (If yes) can you explain how and why? Could you please prioritize the items, telling us why you have chosen the order you have?	To deepen on understanding their conception about research and to know about possible changes ***Then link it to the answers in the research conceptions scale!**

Part III: Journey Plot Instrument (Total Time: About 15–20′)

Question/Theme/Prompt	Notes
• Now I would like to explore the significant events you lived during your doctoral/postdoctoral trajectory. • I would like you to draw a line that represents your trajectory during this last year, with highs (positive significant events) and lows (negative significant events). (*Do a graphic so he/she can see it*).	Explain how the journey plot works Ask the participant to explain the graphic while he/she draws. Use the participant's grid to write those significant events that later will need to relate with the network plot.
For each significant event (*draw a circle on each significant event so the participant sees the event we are talking about*): • Let us look at each of these events in more detail • In this case, what actually happened was…. (*repeat what the interviewee has said before*) • How long did it last? • How did you feel then? And now? How your feelings evolved? • Who was involved in this event? • Who helped you to overcome the situation? How did he/she/they help you? • In looking across all the emotional lows, can you tell me anything about how you coped with these kinds of experiences?	Note them the intensity of the events (e.g., is that event even more positive than this previous one?) **Remember!** Each significant event will have to be related to a certain individual or group in the "network plot"

Question/Theme/Prompt	Notes
General questions: • Looking at the shape of the graphic, how you would describe your overall experience/journey? (if necessary remind her/him some characteristics of the graphic: ***many changes and variations/your graphic is very flat/you started in very high but your experience went down along the journey/...,)***	General questions to make sense and have more information about the general picture of the graphic
Significant events in the questionnaire: • In the questionnaire you explained that **XXX** was one of the most significant positive events and **XXX** one of the most negatives ones. Do they appear in this graphic? (*If not*) When did they happen? • Is there any relation between this one and the ones you experienced this past year? Same questions than for the significant events (what, how long, feelings…).	Use the grid of the interviewee to collect the information that might be useful to prompt recall.

Part IV: Network Plot (Total Time: About 15–20′)

Question/Theme/Prompt	Notes
This is the tool we will use to represent your research network in the scientific community, that is, the individuals and the groups you interact with **Possible prompt if the interviewee does not know how to start:** In the questionnaire you mentioned you did a research stay, you work in a research team, you participated in different conferences….	*Explain how the Network Plot works* Use the grid of the interviewee to collect the information that might be useful to prompt recall.
Why is this individual important for you? Do you feel comfortable with your position in relation to this individual/group? Why? Or why not? Are all of them equally important?	Note interviewee's position in relation to the other circles
Are those events you have mentioned in the journey plot related to some of these individuals and groups (looking at the network Plot)? Which are the relationships that were important for each event in relation to the individuals and groups in your network plot?	Note participants about the relationships between individuals/groups and the sizes they choose.
Where would you like to be this network constellation? What would it take for you to achieve this position?	Ask to the participant to draw another circle with his/her desired position

Part V: Closure (Total Time: About 5')

Question/Theme/Prompt	Notes
Do you want to add anything else?	Opportunity for the interviewee to add or expand on what has been discussed

REFERENCES

Åkerlind, G. (2005). Postdoctoral researchers: Roles, functions and career prospects. *Higher Education Research & Development, 24*(1), 21–40. http://dx.doi.org/10.1080/0729436052000318550.

Åkerlind, G. S. (2008). An academic perspective on research and being a researcher: An integration of the literature. *Studies in Higher Education, 33*(1), 17–31.

Bazerman, C., Keranen, N., & Encinas, F. (2012). Facilitated immersion at a distance in second language scientific writing. In M. Castelló, & C. Donahue (Eds.), *University writing. Selves and texts in academic societies Volume series in writing:* (pp. 235–248). Bingley, UK: Emerald group Publishing Limited.

Bereiter, C., & Scardamalia, M. (1987). *The psychology of written communication*. Hillsdale, NJ: LEA.

Boice, R. (1993). Writing blocks and tacit knowledge. *Journal of Higher Education, 64*(1), 54.

Brew, A. (2001). Conceptions of research: A phenomenographic study. *Studies in Higher Education, 26*(2), 271–285.

Bryman, A. (2006). Integrating quantitative and qualitative research: How is it done? *Qualitative Research, 6*(1), 97–113.

Buckingham, D. (2009). Creative" visual methods in media research: Possibilities, problems, and proposals. *Media Culture Society, 31*(4), 633–652.

Cameron, J., Nairn, K., & Higgins, J. (2009). Demystifying academic writing: Reflections on emotions, know-how and academic identity. *Journal of Geography in Higher Education, 33*(2), 269–284. http://dx.doi.org/10.1080/03098260902734943.

Cantwell, B. (2011). Academic in-sourcing: International postdoctoral employment and new modes of academic production. *Journal of Higher Education Policy and Management, 33*(2), 101–114.

Castelló, M., Iñesta, A., & Corcelles, M. (2013a). Ph. D. students' transitions between academic and scientific writing identity: Learning to write a research article. *Research in the Teaching of English, 47*(4), 442–478 Special Issue.

Castelló, M., Iñesta, A., & Corcelles, M. (2013b). Ph. D. students' transitions between academic and scientific writing identity: Learning to write a research article. *Research in the Teaching of English, 47*(4), 442 Special Issue.

Castelló, M., Kobayashi, S., McGinn, M., Pechar, H., Vekkaila, J., & Wisker, G. (2015). Researcher identity in transition: Signals to identify and manage spheres of activity in a risk-career. *Frontline Learning Research, 3*(3), 39–54.

Castelló, M., McAlpine, L., & Pyhältö, K. (2017). Spanish and UK post-PhD researchers: Writing perceptions, well-being and productivity. *Journal of Higher Education Research & Development*, 1–15. http://dx.doi.org/10.1080/07294360.2017.1296412.

Cobb, S. (1976). Social support as a moderator of life stress. *Psychosomatic Medicine, 5*(38), 300–314.

Corcelles, M., Cano, M., Liesa, E., González-Ocampo, G., & Castelló, M. (2017). Constructing academic trajectories: What is the role of positive and negative events in the doctoral journey? *Journal of Research in Higher Education* (submitted).

Creswell, J., & Plano-Clark, V. (2011). *Designing and conducting mixed methods research.* Thousand Oaks, US: Sage.

Elo, A.-L., Leppänen, A., & Jahkola, A. (2003). Validity of a single-item measure of stress symptoms. Scandinavian Journal of Work. *Environment & Health, 29*(6), 444–451.

Florence, M. K., & Yore, L. D. (2004). Learning to write like a scientist: Coauthoring as an enculturation task. *Journal of Research in Science Teaching, 41*(6), 637–668.

Greene, J. C., Caracelli, V. J., & Graham, W. F. (1989). Toward a conceptual framework for mixed-method evaluation designs. *Educational Evaluation and Policy Analysis, 11*(3), 255–274.

House, J. (1981). *Work stress and social support.* Addison-wesley educational Publishers Inc.

Hyland, K. (2002a). Authority and invisibility: Authorial identity in academic writing. *Journal of Pragmatics, 34*, 1091–1112. http://dx.doi.org/10.1016/S0378-2166(02)00035-8.

Hyland, K. (2002b). Genre: Language, context, and literacy. *Annual Review of Applied Linguistics, 22*, 113–135.

Kamler, B., & Thomson, P. (2008). The failure of dissertation advice books: Toward alternative pedagogies for doctoral writing. *Educational Researcher, 37*(8), 507–514.

Kyvik, S. (2013). The academic researcher role: Enhancing expectations and improved performance. *Higher Education, 65*(4), 525–538.

Laudel, G., & Gläser, J. (2008). From apprentice to colleague: The metamorphosis of early career researchers. *Higher Education, 55*(3), 387–406.

Lea, M. R., & Stierer, B. (2011). Changing academic identities in changing academic workplaces: Learning from academics' everyday professional writing practices. *Teaching in Higher Education, 16*(6), 605–616.

Lonka, K., Chow, A., Keskinen, J., Hakkarainen, K., Sandström, N., & Pyhältö, K. (2014). How to measure PhD students' conceptions of academic writing – and are they related to well-being? *Journal of Writing Research, 5*, 245–269.

Maslach, C., & Leiter, M. P. (2008). Early predictors of job burnout and engagement. *Journal of Applied Psychology, 93*, 498–512. http://dx.doi.org/10.1037/0021-9010.93.3.498.

Maslach, C., Schaufeli, W., & Leiter, P. (2001). Job burnout: New directions in research and intervention. *Current Directions in Psychological Science, 12*, 189–192. http://dx.doi.org/10.1111/1467-8721.01258.

McAlpine, L. (2016). Becoming a PI: Shifting from 'doing' to 'managing' research. *Teaching in Higher Education, 21*(1), 49–63.

McAlpine, L., & Amundsen, C. (2016). *Post-PhD career trajectories: Intentions, decision-making and life aspirations.* Basingstoke: Palgrave Pivot.

McAlpine, L., Amundsen, C., & Turner, G. (2014). Constructing post-PhD careers: Negotiating opportunities and personal goals. *International Journal for Researcher Development, 4*(1), 39–54.

McAlpine, L., Pyhältö, K., & Castelló. (2017). *Building a more robust conception of early career researcher experience: Integrating qualitative and quantitative processes* (submitted).

Miller, N., & Brimicombe, A. (2003). Mapping research journeys across complex terrain with heavy baggage. *Studies in Continuing Education, 26*(3), 405–417.

O'Meara, K., Jaeger, A., Eliason, J., Grantham, A., Cowdery, K., Mitchall, A., et al. (2014). By design: How departments influence graduate student agency in career advancement. *International Journal of Doctoral Studies, 9*, 155–179.

Paré, A., Starke-Meyerring, D., & McAlpine, L. (2009). The dissertation as multi-genre: Many readers, many readings. In C. Bazerman, A. Bonini, & D. Figueiredo (Eds.), *Genre in a changing world* (pp. 179–193). Fort Collins, Colorado: The WAC Clearinghouse and Parlor Press. Available at http://wac.colostate.edu/books/genre/.

Pyhältö, K., McAlpine, L., Peltonen, J., & Castelló, M. (2017). How does social support contribute to engaging post-PhD experience? *European Journal of Higher Education*, in press. https://doi.org/10.1080/21568235.2017.1348239.

Pyhältö, K., Nummenmaa, A.-R., Soini, T., Stubb, J., & Lonka, K. (2012). Research on scholarly communities and the development of scholarly identity in Finnish doctoral education. In S. Ahola, & D. M. Hoffman (Eds.), *Higher education research in Finland* (pp. 337–354). Jyväskylä: University of Jyväskylä.

Pyhältö, K. M., Peltonen, J., Rautio, P., Haverinen, K., Laatikainen, M., & Vekkaila, J. E. (2016). *Summary report on doctoral experience in UniOGS graduate school at the University of Oulu*. Acta Universitatis Ouluensis. Retrieved from http://jultika.oulu.fi/Record/isbn978-952-62-1084-1.

Pyhältö, K., Vekkaila, J., & Keskinen, J. (2015). Fit matters in the supervisory relationship: Doctoral students' and supervisors' perceptions about the supervisory activities. *Innovations in Education and Teaching International*, *52*(1), 4–16. http://dx.doi.org/10.1080/14703297.2014.981836.

Reybold, E., & Alamia, J. (2008). Academic transitions in Education: A developmental perspective of women faculty experiences. *Journal of Career Development*, *35*(2), 107–128.

Rose, M. (1980). Rigid rules, inflexible plans, and the stifling of language: A cognitivist analysis of writer's block. *College Composition and Communication*, *31*(4), 389–400.

Ross, R. G., Greco-Sanders, L., & Laudenslager, M. (2016). An institutional postdoctoral research training Program: Increasing productivity of postdoctoral Trainees. *Academic Psychiatry*, *40*(2), 207–212.

Ross, R. G., Greco-Sanders, L., Laudenslager, M., & Reite, M. (2009). An institutional postdoctoral research training program: predictors of publication rate and federal funding success of its graduates. *Academic Psychiatry*, *33*(3), 234–240.

Russell, D. R., & Cortes, V. (2012). Academic and scientific texts: The same or different communities. In M. Castelló, & C. Donahue (Eds.), *University writing. Selves and texts in academic societies Volume series in writing:* (pp. 3–18). Bingley, UK: Emerald group Publishing Limited.

Sala-Bubaré, A., & Castelló, M. (2017). Exploring the relationship between doctoral students' experiences and research community positioning. *Studies in Continuing Education*, *39*(1), 16–34.

Sawyer, K. (2009). Writing as a collaborative act. In S. B. Kaufman, & J. C. Kaufman (Eds.), *The psychology of creative writing* (pp. 166–179). Cambridge: Cambridge University Press.

Starke-Meyerring, D., Paré, A., Artemeva, N., Horne, M., & Yousoubova, L. (Eds.). (2011). *Writing (in) knowledge societies*. West Lafayette: Parlor and Fort Collins: The WAC Clearinghouse. Retrieved from http://wac.colostate.edu/.

Torrance, M., Thomas, G. V., & Robinson, E. J. (1994). The writing strategies of graduate research students in the social sciences. *Higher Education*, *27*(3), 379–392.

Väisänen, S., Pietarinen, J., Pyhältö, K., Toom, A., & Soini, T. (2016). Social support as a contributor to student teachers' experienced well-being. *Research Papers in Education*, 1–15. http://dx.doi.org/10.1080/02671522.2015.1129643.

Veikkaila, J., Virtanen, V., Taina, J., & Pyhältö, K. (2016). The function of social support in engaging and disengaging experiences among post PhD researchers in STEM disciplines. *Studies in Higher Education*, *5*, 222–235. http://dx.doi.org/10.5430/ijhe.v5n2p222.

Vekkaila, J., Pyhältö, K., & Lonka, K. (2013). Experiences of disengagement – A study of doctoral students in the behavioral sciences. *International Journal of Doctoral Studies*, *8 (2013)*, 61–81.

Vekkaila, J., Pyhältö, K., & Lonka, K. (2014). Engaging and disengaging doctoral experiences in the behavioural sciences. *International Journal for Researcher Development*, *5*(1), 33–55. http://dx.doi.org/10.1108/IJRD-09-2013-0015.

Woehrer, V. (2014). To say or to go? Narratives of early-stage sociologists about persisting in academia. *Higher Education Policy*, *27*, 469–487.

Chapter 8

Postdoc Trajectories: Making Visible the Invisible

Lynn McAlpine[1,2]
[1]*University of Oxford, Oxford, United Kingdom;* [2]*McGill University, Montreal, QC, Canada*

Chapter Outline

Goal of Chapter	**176**
Context	**177**
The Evidence I Am Drawing on	**178**
Study 1 (2006–2016)	179
Study 2 (2014–2015)	180
Study 3 (2015–2016)	180
Study 4 (2013)	181
What Meaning Do I Draw From the Research?	**182**
What Do Postdocs Actually Do?	182
Categories of Work: So, What Is in a Name?	183
Agency, Motivation, Resilience: Variation and Its Implications	185
The Personal: Interaction With Work	187
How Does Their Work Contribute to Building Their Hoped-for Careers?	189
Mobility: Role in Career Trajectory	189
Policy Knowledge: Responsibilities and Resources	190
Where Do They Foresee Themselves in 5 Years?	**193**
Still Hope	193
Considering Alternatives	194
Implications: What Might All This Mean for the Future?	**195**
Appendix A	**198**
References	**201**

The Postdoc Landscape. http://dx.doi.org/10.1016/B978-0-12-813169-5.00008-2

Otto, a Scientist, Now in the United Kingdom

Otto completed his education including a PhD in the United Kingdom and then moved to Australasia where he held two successive postdoc contracts. The transition to postdoc was challenging, particularly in terms of "getting everything set up in a new environment and taking on a lot more challenges." He published several times in good journals, which was "a huge high." The real problem was effectively managing his own research alongside his responsibilities to the principal investigator whose projects he was funded on. At one point, he was fortunate to get an internal grant of his own, which allowed a "bit more freedom to follow your own nose," though it was difficult to juggle the two workloads. At that point, he began looking elsewhere for work, given the "distinct lack of criteria and guidelines" for getting permanent status in either of his positions. Ultimately, he had two choices: stay in Australasia on a semiindependent grant as a senior researcher in an established group or move to the United Kingdom on a start-up fellowship, which meant he would have to procure funding within 5 years. There was a degree of risk, especially given that he had a family, but he moved back to the United Kingdom. Fortunately, 9 months later, he got his first grant, 5 years after graduating.

Romeo, a Scientist, Now in Europe

Romeo moved from Europe to North America to pursue his PhD. After completing the degree, he applied for 40–50 postdocs, but discovered the positions he got often resulted from colleagues' recommendations rather than blind applications: "You only get offers anyway from places where people know you." In the next few years, he did postdocs, often short, in the United States, Latin America, two places in Canada, and the United Kingdom. In his first postdoc, he presented a paper from his thesis, which drew attention and led to a still-cited paper. He returned home when a senior colleague wrote a fellowship application with him in mind. Once he returned, he applied for a new researcher grant but was unsuccessful the first time; he was happy to be successful on the second try (his last opportunity). The grant funded his salary, but without funding for a team. Thereafter, he applied for the next level of grant, feeling pressured to succeed since getting the grant would help with gaining permanence. There was no guarantee of success, which made him feel demotivated. He even thought of giving up research if he did not get the grant, since failure would mean he "was not worth the money." Thankfully, he was successful, 7 years after graduating.

Excerpts adapted from McAlpine, L., 2016a. Becoming a PI: Shifting from 'doing' to 'managing' research. *Teaching in Higher Education,* 21, 49–63) and McAlpine, L., Turner, G., Saunders, S., & Wilson, N. 2016. Becoming a PI: Agency, persistence, and some luck!. *International Journal for Researcher Development,* 7, 106–122.

GOAL OF CHAPTER

The cameos above (data collected in 2014–2015) focus on two individuals' efforts to obtain a funding grant to become a principal investigator (PI) and provide a glimpse into the lives of postdocs today. Their experiences and those of their colleagues are the focus of this chapter. What do postdocs actually do? How does their work contribute to building their career goals? And where do they foresee themselves in 5 years? This chapter explores these questions by drawing on over a decade of research into the experience of postdocs across a range of disciplines in Canada, the United Kingdom, and continental Europe. It locates their career hopes and aspirations within the broader contexts of research policy and funding as well as institutional commitment to career development. The hope is to make visible a number of factors influencing postdoc career trajectories that are often invisible in the debate over, consideration of, and adoption of national policies, institutional procedures, and organizational human resource offerings.

CONTEXT

Internationally, the career trajectories of postdocs are precarious (Powell, 2015). They often find themselves caught between a passionate personal investment in their research and work environments, which often offer little future, given soft research funding as well as institutional contingent employment structures (McAlpine, 2010). This situation is neither new nor, unfortunately, is it improving; it appears, indeed, to be worsening, partly due to the fallout of the global economic crisis, which several research participants referenced. Say, a scientist, describes this "Catch 22" extremely well (stated in 2013):

> I didn't really want a conditional [postdoc] appointment [but] with the economy the way it is …universities put hiring freezes in so … there are fewer assistant professor jobs. There is this glut of post-docs now …an infinite supply … And so, you know, the PIs have had the upper hand …like they can essentially just say to you, "Well I got 30 other people who emailed me this week." … It is …becoming a …systemic problem … Someone has to take leadership …universities need to limit lengths of post-docs or …limit the number of PhDs they award or …the system needs to be controlled in some way so that …we don't over train PhDs and then have them expecting jobs when those jobs don't exist …universities want to train as many as possible, but in my opinion …that's not really all that reasonable when there is a massive glut of people available … So the job prospects are not extremely bright.

What Say points to is how policies or lack of policies (sometimes influenced by global events) affect the possibilities for postdocs: PhD graduates who wish to remain in the academy, find secure work, develop their research profiles, and gain tenure-track positions.[1] Say focuses largely on the institutional context, but this context is nested within others (McAlpine & Norton, 2006), which are equally influential (e.g., national level: the amount of funding available from

---

1. Throughout this chapter, North American terms are used.

funding councils and how they choose to structure the grants they are offering: blue skies or strategic; international: EU policies). I will return to this issue of policies later.

What is also hidden in Say's comment is the way in which academic work is changing. There is greater specialization within the academic workforce; for instance, the growth of teaching-only positions—which are often more secure than postdoc positions since they are funded institutionally rather than through grants—represents one big change. There is greater casualization where individuals are (re-)employed on a casual or short-term basis, and tenure has largely disappeared in some English-speaking countries. In addition, it is not unusual in some places for professional staff to perform aspects of teaching and research functions (e.g., program directors). Concurrently, the development of new technologies has shifted the nature of research-related work dramatically (e.g., virtual data collection, ever larger data sets, mass computing). It has also attenuated the value of the physical university as a socially powerful workspace, with individuals no longer going to the university except for specific purposes (e.g., a meeting) (McAlpine & Mitra, 2015). Within this context, I turn now to what our team research has helped us understand about postdoc experience: first by overviewing the evidence I will draw on, and then by exploring the meanings I draw from the different analyses.

THE EVIDENCE I AM DRAWING ON

Along with teams in Canada and the United Kingdom, I have been researching early career researcher experiences since 2006, using largely qualitative methods. By "early career researchers," we mean doctoral students and those in the first few years post-PhD, regardless of the kinds of work they do. Our methodological stance is generally a narrative one (Elliott, 2005), meaning we focus first on the totality of the individual's account (whether textual or oral), the story that the individual has recounted that represents part of his or her life history. It is only within this broader context that we then begin to think about the specific themes or patterns that may emerge across individuals. This is why I introduced the chapter by offering two postdocs' accounts of their lives (or parts of their lives) since the themes I will discuss can often be better understood as specific examples within individuals' lives, rather than by disembodied quotes. I will introduce six other short narratives, cameos, in the course of this chapter and draw principally on the eight accounts to ground the themes I discuss. Lastly, my conceptual lens is strongly influenced by the notion of identity development[2] (McAlpine, Amundsen, & Turner, 2014),

---

2. We are not alone in our interest in identity development; see gender studies (Sondergaard, 2005) and psychology (Côté , 2005).

which is the way individuals make sense of and learn from their experiences as they construct and reconstruct the meaning of their life experiences. Below, I describe the four different studies I am drawing on: in each case, describing the data and the general findings as a context for my answers to the three questions structuring the chapter.

Study 1 (2006–2016)

In this longitudinal study, we followed 48 scientists and social scientists from 5 to 7 years between 2006 and 2016 with the goal of understanding how their day-to-day experiences influenced their career hopes and decisions. They were recruited in Canada and the United Kingdom, but at the end of the study were distributed globally across 30 institutions. Of the 48, 24 had experience being postdocs and a number still remained postdocs at the end of the study. It is the 24 postdocs' stories I will draw on: 9 had achieved a tenure-track position, 12 postdocs still remained hopeful, and 3 were considering other options (see Appendix A, Table 8.1 for demographic information).

Each participant provided a range of data types on an annual basis (biographic questionnaire, weekly activity logs, preinterview questionnaire, and interview, sometimes using visual methods[3]). For all individuals, I am drawing on at least 20 different data collection instruments, sometimes as many as 30. It was this work that led to the construct of identity-trajectory, which provides a unique way of examining career decision-making and development, with career understood as an individual's course or progress through life or a part of life (McAlpine & Amundsen, 2016); in other words, the notion of "career" also incorporates personal goals (i.e., life tasks, quality of life) as well as paid employment—a broader definition than that often used in day-to-day discourse.

Identity-trajectory incorporates a view that the past influences the present and the future, so there is a continuity of stable personhood over time (individuals remain who they are, regardless of age, experience, etc.), while at the same time experiencing a sense of ongoing change (being different from 6 months ago, such as more accepting of contrary views) (McAlpine et al., 2014). The notion of identity-trajectory that emerged relatively early on in the research led to a focus on how individuals perceive and respond emotionally to their experiences, their sense of personal agency (with agency intertwined with emotion), and how learning from both emotionally positive and negative experiences contributes to new ways of thinking and acting. We also learned from

---

3. Visual methods involve the use of images, whether provided by the interviewer, created by the interviewee, or a combination of the two. Visual methods can often capture a less inhibited, more spontaneous account of experience than interviews alone (Bagnoli, 2009).

this research that work and workplace decisions need to be situated within the individual's broader personal life, hopes, and intentions.

Study 2 (2014–2015)

The second qualitative study included 60 postdocs from all disciplines in the United Kingdom and Europe (see Appendix A, Table 8.2 for demographic information). In this study, we examined the journey from PhD graduation to getting the first grant as PI role, and, secondly, what being a PI meant. The data for this study, collected in 2014–2015, included a biographic questionnaire, a CV, and an interview using another visual method (McAlpine, 2016a; McAlpine, Turner, Saunders, & Wilson, 2016).

What was striking was the nuanced perspective the PIs provided on the interaction between individual agency and funding as well as career environments where the "randomness factor" played a role in success. In such an environment, agency and resilience, the ability to manage one's emotional response to experiences and sustain self-belief, were important to continuing the journey to getting a grant. In fact, their key advice was best summed up as "drive your own success." Once having gained a grant, individuals experienced new challenges they were unprepared for they found themselves distanced from what they loved, actively researching. Aside from these findings, we were also struck by several other factors: the degree of mobility, the view that international experience was critical, and the length of time (up to 11 years) that it took for some to become PIs.

Study 3 (2015–2016)

This analysis draws on a subset of data from a large mixed-method study (see Chapter 6, Castello, Pyhalto, and McAlpine) examining the experiences and difficulties that doctoral students and postdocs from a range of disciplines and in different European countries face in constructing their scholarly identities.[4] All participants completed a survey, and then a subset were interviewed, again including visual methods. The focus of the mixed-method analysis below is data from 71 postdocs in the United Kingdom: 71 surveys and interviews with a subset of 10 of the 71 (see Appendix A, Table 8.3). The data were collected in late 2015–early 2016, so analysis is ongoing.

We were interested on the meaning or sense that postdocs made of experiences they characterized as emotionally powerful since graduation

---

4. The study I describe focuses on the emotional meaning of significant events experienced by postdocs, whereas the next chapter provides a broader perspective on the overall research program and looks at how participants with different writing profiles deal with experiences involving writing.

(Skakni and McAlpine, in press). We began with the survey analysis and then turned to the interviews to ground and contextualize the survey results. In the survey analysis, individuals reported a range of precipitating significant events, with emphasis on getting published, getting a job, and getting a grant. We found that the meanings of the positive and negative events could be characterized in one of two ways: either as a direct effect/consequence of the outcome of the event or as broader learning (lessons learned) for the future. What was striking was that lessons learned for the future *only emerged from negative experiences*, and never from positive emotional experiences. We conjecture that negative emotional experiences drew forth greater reflection and thus learning as they challenged views of self, whereas positive emotional experiences were confirmatory. The analysis of the interviews demonstrated that such high and low emotional moments occurred relatively frequently (3–5 times a year), and that a precipitating experience could incorporate both high and low moments.

Study 4 (2013)

In this last study, we explored how early career researchers understood, responded to, and negotiated their careers in relation to higher education policies since policy strongly influences indeed, in some cases, drives the nature of academic work. In 2013, we interviewed 42 early career researchers from a range of countries (13, 31%, of whom were postdocs), all in the field of higher education (Ashwin, Deem, & McAlpine, 2015) (see Appendix A, Table 8.4 for demographic information). We chose individuals researching in the field of higher education thinking that they might be more likely to be knowledgeable about policy influences on their work than those researching in other fields. What we found, in fact, was little awareness of policy. Further, many participants perceived the relationship between themselves and higher education policies from an individual perspective rather than in the broader context of a community of researchers. Still, those with more experience, whom we called policy "actors," generally tended to report being more agentive in negotiating the influence of policy on their experiences, providing concrete examples of what they did. Those with less experience tended to feel unable to negotiate the influence of policy on their work which as you will see puts them at a disadvantage in navigating their careers.

I hope the brief descriptions above provide a sense of the broad data set—over 100 postdocs representing all disciplines and situated in a range of European countries, the United Kingdom, and North America— that I am drawing on in addressing these questions: What do postdocs actually do? How does their work contribute to their career goals? And where do they foresee themselves in 5 years? Before starting though, here are two more cameos, drawn from Study 3.

Rob, 2nd Year Postgraduation, 42, Social Scientist, Now in the United Kingdom

Rob had just earned a tenure-track position in a teaching-focused university after a year of part-time postdoc contracts. He hoped to build his reputation there and apply later to a research-intensive university "with...more publications and...research experience behind me." He had two children.

Rob enjoyed his work, felt respected by his community, and experienced a sense of collegiality. He also received encouragement, personal attention from his supervisor, and felt appreciated.

Rob reported many multiinstitutional connections with individuals. He made a lot of "really good...connections" at conferences, so his network was "geographically...very scattered...in the United Kingdom or mainland Europe." He remained connected with people where he did his PhD (including his supervisor). He worked to build his network, for instance, working with present university colleagues not in his field to get internal research funding. He also had "done research with [researchers elsewhere]...and met a lot of people...that's really important in...my career." He was "getting a really good picture of...who the big players are."

Despite work commitments, for Rob, "family is ...central...[still] I am the only earner in the family," so his childcare responsibilities were mostly weekend-related since "my wife needs a break." Also, he "thought about leaving the project given the lack of security [but only] if a better job came along." Unlike Sandra, Rob did not report exhaustion.

Sandra, 5th Year Postgraduation, 36, Scientist, Now in the United Kingdom

Sandra had first held a postdoc fellowship, before a number of contracts with the same PI. She felt the "pressure to publish...takes away...the enthusiasm...I felt about doing research" initially. Last year, she chose not to renew her contract—making her dependent on her partner. She had two children.

Like Rob, Sandra enjoyed her work and felt appreciated by and collegial with her research community. Her supervisor gave her personal attention and encouraged her; she felt appreciated.

Sandra described many multiinstitutional and international connections in "our niche [post-PhD] area" including her present supervisor: two research groups in the United Kingdom, three in Europe and one in North America. The two European groups were most important: "We've done a lot of work together and we exchange a lot of...ideas and... have a lot of discussions." Like Rob, Sandra could "contribute and interact," suggesting a supportive research community. But, she said, "If you disconnect me from my supervisor, I'm...outside the group," suggesting she did not see herself as initiating these relationships.

Sandra reported relatively high levels of exhaustion: "I have to combine the responsibilities of having a young family with long hours required for work...I...feel that I'm not able to achieve what's expected from work" [and] "there is...[the] expectancy to sacrifice personal life. [It's] difficult to combine work and children."

Excerpt adapted from McAlpine, L., Pyhalto, K., & Castello, M. (under review). Building a more robust conception of early career researcher experience: what might we be overlooking? *Studies in Continuing Education.*

WHAT MEANING DO I DRAW FROM THE RESEARCH?

What Do Postdocs Actually Do?

The post-PhD period is one of transition from the PhD period where individuals frequently focus on finishing. This time is generally one in which individuals are trying to create their own niche and enhance their job potential through publications (recall Rob's plan to publish to be competitive), networking for support (recall Romeo's comment about knowing others), and collaboration (Sandra's interaction with her international colleagues), and hopefully secure research funding.

What I am interested in is the nature of postdocs' work experiences, whether as scientists or social scientists. In highlighting the invisible, I look first at the varied roles and responsibilities of "postdoc" work, which may not actually involve much research despite the relatively common assumption that this is what postdocs do. I then turn to individual agency: the variation that was evident in how individuals invested in advancing their careers and maintained motivation for their work. Finally, I consider how their broader personal lives interacted with their work actions and decisions.

Categories of Work: So, What Is in a Name?

The word "postdoc"[5] has been defined as study or work undertaken after the receipt of the doctorate. Such a definition incorporates two potentially distinct tasks and fails to differentiate between those funded on fellowships (independent and personally directed learning) and those funded through contracts with PIs (responsible for carrying out designated work). We cannot—in fact, should not—ignore the difference in sources of funding since the difference had fundamental effects on work and motivation.

Those who held fellowships at various points after PhD graduation enjoyed the opportunity to follow their own intellectual interests and concurrently develop their research profiles and careers—if only for a defined period of time. Nevertheless, fellowships came in different forms and lengths, so they had different effects on participants' career paths. Brookeye, for instance, held a 2-year fellowship immediately after graduation. He had a choice of where to work and could decide what involvement he had with those where he was located. Initially he focused on getting a job to establish his future financial security. Then, he directed all his attention to building his profile, putting off teaching and supervision until his tenure-track position, and used the time to visit scholars elsewhere and build collaborations. He also decided to. However, such fellowships were not always full-time. Otto, whose story you read at the beginning of the chapter, also received a fellowship, but it was only part-time. So, while he had the freedom part-time to follow his own interests, he was doing this while still juggling contract work. Others, like Catherine, obtained a 1-year fellowship after several years as a contract researcher. She thoroughly enjoyed the release from her institutional responsibilities, if only for a year, and like Brookeye and Otto, she used the time to develop collaborations and publish. Still, after the fellowship,

5. In Europe and the United Kingdom there has been a tendency to refer to those with PhD degrees as "researchers" rather than postdocs.

she returned to contract work and was not able to sustain the momentum she had built up during the year.

On the other hand, those working on contracts not only dealt with financial insecurity but also often part-time contracts such as Rob's (cited above). Individuals reported they often found themselves in tension between doing the work required of them and their own "passionate investment" (McAlpine, 2010). Individuals were trying to maintain their own intellectual work (and networking) while being paid to do work often outside of or on the periphery of their own interests and expertise. A new contract could involve an "intellectual relocation," an investment of time and energy to "come up to speed" on research they might be unfamiliar with. Thus, making a contribution was difficult because these individuals did not necessarily know the discourse, the ideas, and the thinkers important to their new colleague. So much time had to be invested to feel competent in the new area (knowing what the important questions, knowing where to publish) to contribute. This could be a serious handicap since individuals need to be able to contribute and publish to progress their careers. While many tried to maintain their own research interests, not all did; and like Sandra, they became more focused on the PI's work, given other influences on their lives.

Further, being on a research contract need not necessarily involve research. Margarida held year-by-year contracts in three different universities and characterized research as only 20% of her work; teaching took up 30% and management occupied 50% of her time. Each year she waited to hear if the contracts would be renewed. Marian reported she did mostly teaching. Lastly, while many postdocs worked in positive environments, some reported themselves doing, as Charles said, "scut work," which made the work even less appealing. Similar disappointments were experienced by Anne, who worked in an unorganized lab, who found she was often doing what the lab manager should, and Funky Monkey, who found his postdoc "worse than grad school," dealing with difficult and failed experiments that made him question the meaningfulness of the research. This was not only demotivating but also made it difficult to get publications, which were necessary to be competitive.

Generally, individuals on contract reported they felt treated like second-class citizens. For instance, Fracatun and Onova wanted to teach, but their departmental policy prevented them from doing so. Others noted that they could not access training or conference funding. Even things as straightforward as receiving departmental emails and having mail slots could not be taken for granted, as Jennifer noted.

Those who achieved their long-term goal of being awarded a grant as PI found a different set of challenges, which as Juliet remarked involved "not doing what I thought I would be doing." Despite their love of research, they discovered they were now managing, not doing, research. These management tasks were ones they felt they had not been prepared for financially overseeing the grant, managing a team (including dealing with "personnel" issues), negotiating the "political" environment—overall developing a management style that would enable them to get people to do things they did not always want to do.

Finally, their work patterns, regardless of source of income, were influenced by new technologies. Scientists, for instance, reported doing research-related work "anywhere where I can find a seat, planes trains and automobiles, coffee shops and cafes…bathrooms" (AAA). Flora's "list of favorites" included

> *Home (kitchen, bedroom, balcony), metro, field work (southern US, northern Canada), hospital (cafeteria and hospital bedroom – while visiting my mom hospitalized), train, airport, city park, my parent's house (kitchen), hotel room.*

Such flexibility could be helpful. As Anne noted, "Grant and paper writing, abstract writing can take place anywhere and it often saves commuting time to do these at or near home." Still, it could also intrude as individuals might never feel off-duty. Indeed, many commented on the lack of work–life balance.

Agency, Motivation, Resilience: Variation and Its Implications

What should be apparent already is that individuals' love of research helped sustain their motivation. Still, there was considerable variation across individuals in how they undertook to direct and negotiate their work environment and their future careers, a capability we refer to in our research as agency. By agency, we mean individuals' efforts to be intentional, to plan, to construct a way forward toward their goals, whether work or personal (McAlpine & Amundsen, 2016).[6] Agency had been well documented in the literature for the past several decades. Agentic efforts include being flexible and adapting to constraints (whether expected or unexpected) and trying to negotiate the constraints, though not always successfully. Agency offers a way to understand investment in work and, further, how work decisions are embedded within personal goals, challenges, relationships, and responsibilities (see next section).

Generally, individuals varied in the degree to which they reported or demonstrated efforts to be agentive at work (e.g., planning, organizing, knowing the specifics of what was needed, and working to make it happen). Dancer exemplifies a strong sense of agency despite a challenging work context. She was no longer on contract, and was, in fact, unemployed when interviewed, yet she continued to engage in work activities:

> *I attend the [team] meetings from time to time. I'm still producing publications, and will probably continue for…for several years to come …I'm trying to make it an opportunity …I'm still working full-time on my own on these publications, I'm still going fieldwork, I'm going to conferences, so I'm keeping myself active in exactly the same way as I was doing [when paid], except that I don't have an income so…that's the only difference.*

---

6. Others also share an interest in agency: see Chapter 3. We draw particularly on the work of Margaret Archer (2003).

Those who exhibited and sometimes named their lack of agency as a handicap appeared, at least in certain contexts, to demonstrate the following behaviors: to procrastinate, feel unable to act, or depend on external deadlines to undertake or complete tasks. Others, like Simon, who desired a secure future, did not seem to set goals and act on them to advance this goal.

There were many aspects of postdoc work that could be approached from an intentional strategic perspective. For instance, Brookeye and Flora were very careful in their choice of institution and postdoc supervisor; they attended to name recognition of their supervisor, potential collaborations, and shifts in the growth of research fields. In contrast, others often only learned this later in looking back at their experiences. For example, Anne, who ultimately decided she would not continue to seek a tenure-track position, realized only later that her research difficulties resulted from not having engaged strategically in such decision-making before taking on her position; this meant she had found herself in a malfunctioning work environment, which reduced her research potential. In other words, this variation in agency can be partly understood in relation to individuals' work environments and the extent of their collegial networks, but sometimes it was the lack of forward thinking that precipitated unfortunate work environments.

This same variation in agency could be seen in individuals' efforts to build their networks; later sections of this analysis show that Rob was active in developing his network whereas Sandra tended to depend on the networks which the PI pulled together. The same was true of publishing. Some, like Nathan, worked hard to get published and described the significant negative impact on him when he received multiple rejections in a short-time period; he said, "I felt like I was working too hard to face so much failure." Others, like Charles, put off publishing and realized only later that this put them at a disadvantage. Other contexts in which individuals reported rejections and needing to maintain their motivation included seeking research funding and looking for jobs (recall Romeo's multiple job applications).

Overall, individuals experienced many negatively emotional experiences. We learned two things about how they learned to cope effectively with such potentially damaging psychological reactions. First, many appeared to approach negative emotional responses as learning opportunities for the future, so turned negative experiences into possibly positive future uses. Second, many developed resilience in response to negative experiences, i.e., a positive emotional response to stress.[7] For instance, having coping strategies to deal with negative emotions helped maintain motivation (McAlpine et al., 2016). Still, not surprisingly, some became worn down over time by the cumulative challenges over which they had little control (e.g., poor lab facilities, inability to get published).

---

7. Others interested in stress experienced by early career academics include Bazeley (1999), and Baruch (2004); see also Weick, Sutcliffe, and Obstfeld (2005) et al.

What was, indeed, amazing was the extent to which many sustained their self-belief, resilience, and motivation. What was particularly notable was the extent to which individuals invoked the role of luck in succeeding or not. For instance, Say commented, "The thing that surprises me is how much luck can be involved." While this may seem counterintuitive, such a stance was needed to maintain motivation, to remain agentive, to do all one could do to succeed. Luck provided a means to acknowledge the influences on the outcomes of their efforts that were beyond their control (a view supported by empirical evidence: van Arensburger & van den Besselaar, 2012). So, not succeeding did not mean that they did not deserve to succeed.

The Personal: Interaction With Work

> **Funky Monkey, Scientist: 4 Years Post-PhD to…? Now in Canada**
> *Work context*
> Funky Monkey took 6.5 years to finish his PhD due to trouble with his experiments. When he finished, he did not feel academically competitive, so he looked for a career in industry. However, he was unclear about the expectations or options and did not find anything. So, he took a 3-year postdoc contract in another university. The contract was renewed, but then cut short. As our research ended, he was job searching outside academia.
> *Personal*
> Funky Monkey noted his *health* had been affected by his PhD struggles. He also had had *little social life* and sought out volunteer work to stay connected. Afterward, he got married and wanted *financial security*. He was *limited in his job search* since his partner did not want to move. Then, they *had a* baby, which changed his priorities: he tried to limit his work time. Overall, he regretted the loss of health and stamina he had experienced.
>
> **Sophia, scientist: 2 years Post-PhD to Tenure Track, Now in Canada**
> *Work context*
> Sophia explored PhD options and used her network before applying. She left Europe for Canada (new language) to complete the degree, which she finished in a timely fashion. She then did a short postdoc contract in her PhD supervisor's lab before moving to another country, where she found another postdoc contract. However, she shortly returned to Canada when she received an offer of a tenure-track position.
> *Personal*
> Sophia was willing to move anywhere, but after her PhD, she *waited for her partner to* finish. Since he had received a prestigious postdoc in another country, *she moved countries with him and had a baby* while seeking a research job. When she was offered the tenure-track position, *her partner said "it's your turn,"* so they returned to Canada. She had just had a baby so found it quite stressful to manage her new work responsibilities alongside childcare.

Excerpt Adapted from McAlpine, L., 2016c. Navigating careers: Perceptions of post-PhD researchers. In: *Paper presented at the higher education conference, Amsterdam, Netherlands.*

As you can see in Funky Monkey's and Sophia's cameos, there is a more "private" life story (note underlined words) that runs alongside a more "public" work story. This private story, while not so visible, interacts with individuals' investment in the workplace and career decisions. Further, we saw earlier that Sandra gave up her postdoc work due to tensions between caring for her children and her work. As well, Otto, given his family situation, recognized the particular risk he was taking in moving countries regarding financial security. Individuals reported a range of ways in which the personal interacted with work:

- Work–life balance refers to individuals perceiving work as extending into or intruding into personal life, with many finding it difficult to create boundaries. Flora was rare: despite the pressure to stay in the lab for long hours, she chose to work fewer hours than the others in her research group to ensure a good quality of life.
- Well-being encompasses eating and sleeping well, getting exercise and having a social life. Long term, lack of well-being can lead to stress, anxiety, and burnout. Funky Monkey felt burned out due to his PhD. Chronic illness also had an influence; Jennifer and Elizabeth reported their illnesses dramatically reduced their work stamina.
- Family responsibilities and priorities: Having children, elderly parents or other extended family responsibilities influenced work and career decisions. For instance, Funky Monkey chose not to move because of his partner. Similarly, Argyle wished to invest in family life, so he decided he would not move cities in search of work. Further issues around colocation occurred when both partners wanted tenure-track positions near each other. For instance, Thor Bear turned down more than one tenure-track position since his partner could not get a similar position. A final issue was the priority placed on having children. Men such as Funky Monkey and Brookeye rethought their priorities when their first child arrived, and more than one male postdoc took parental leave. As for women, while some wondered if having children would be "suicidal" for an academic career, they were still prepared to go ahead.
- Personal values represent an individual's conceptions of what is good, beneficial, important, etc. These values could create tension with work. For instance, after graduation, George took a postdoc that gave him the opportunity to apply his knowledge to a cutting-edge field. But he was conflicted and also challenged by his partner and friends for the proprietary values implicit in the research focus. He explained, "It has been enriching [but]…I knew I was trading off some things."
- Life goals represent individuals' hopes and intentions for their lives, which were sometimes in tension with work and personal lives. For instance, Flora wanted to move to seek new experiences and more varied job opportunities, and her partner was not willing to move: "So…if he is not willing…I have to take a decision…like family or career so we'll see."
- Financial duress refers to stress related to financial insecurity, such as salary coming to an end, building up debt, not having decent housing, and not being able to provide what individuals wanted for their children. This sometimes led to profound changes in career intentions. AAA decided to start looking for an

industry job since he no longer wanted his financial insecurity to influence what activities he could afford for his children.

Overall, what emerges from this research is evidence of invisible features of work and life that are influential in work motivation, career thinking, and trajectories. First of all, there is the variation in kinds of work that individuals reported doing: having a postdoc did not necessarily mean principally doing research. Further, the nature of work was influenced by more than whether the work was salaried or emerged through a fellowship. For instance, holding a grant as a PI could also mean less involvement in direct research. As well, variation in the degree to which individuals felt agentive, including managing negative emotional experiences, and feeling able to influence their work environments to achieve their career goals, had powerful influences on their motivation and futures. Lastly, many aspects of the personal, revealed in their more private stories, were powerful drivers regarding individuals' work decisions. Agency and the personal have not often been the focus of attention, but in my view, they are central to any efforts to provide better support to postdocs since they influence so greatly their horizons for action—and the careers they imagine for themselves.

How Does Their Work Contribute to Building Their Hoped-for Careers?

Here I explore the ways in which the broader structural contexts interacted with postdoc experience: beginning with mobility and then moving to policy to describe the ways in which individuals perceived these as influencing their work and their understanding of the career opportunity structures open to them.

Mobility: Role in Career Trajectory

There is a growing literature that suggests mobility can be productive for postdoc career trajectories, what Pedersen (2014) calls horizontal mobility, or movement across higher education institutions rather than across labor sectors. Such mobility provides opportunities to develop networks and potential collaborations, expand expertise, and grow a publishing record. In other words, it appears that mobility may be influential in opening up opportunities, more choices. Mobility may be short or long term. For most, it was a longer-term process, resulting from accepting a contract, but a few participants, especially those with fellowships, could arrange short visits or internships, as Romeo did. Further, while any mobility is good, it was clear that individuals particularly valued international experience and felt it made them more competitive. In looking across all the individuals who have participated in the research studies I am drawing on, there were many stories of frequent mobility—though there were challenges in doing so, in terms of work, their personal lives and those of their families. In other words, mobility carried a toll.

So, while mobility across institutions and national boundaries demonstrated adaptability and hopefully would support individuals' career development, mobility engendered relocations, or disruptions, which also had social and psychological implications for both work and personal lives (McAlpine, 2012). These relocations related to geographical and institutional shifts and sometimes jurisdictional, cultural, and linguistic ones as well. As a result, it is not surprising that it could take years to "fit in" as Sandra notes here:

> *I just hadn't adjusted and I was expecting it to be closer to home [culturally] and it wasn't. I felt I couldn't express how to ask for help and when I tried that, I wasn't heard, which is frustrating after a while.*

Linguistic relocation, particularly, was a challenge, and had profound effects not just for work, but for life generally, including partners and families getting jobs and being happy, so concern for others' adjustment often became paramount to postdocs' own adjustments.

Also, in moving countries, individuals had to learn how to effectively negotiate new funding regimes and research genres. For instance, there are different modes of research funding in the European and North American contexts (Cantwell, 2011), which sometimes took individuals by surprise. Thus, in moving from North America to Europe, Fracatun found that the funding model led to shared lab spaces, which he viewed positively, but that to become permanent a part of his salary would ultimately be dependent on the research funding he was awarded, a challenging discovery when he learned this after being in this new position some time.

Institutional moves even within the same jurisdiction required learning new organizational structures and enabled organizational comparisons, so individuals became much more aware of variation in institutional pressures and departmental politics. This was, in reality, an area where there were some surprises when individuals found out that their assumptions about institutional resources and responsibilities based on their previous institution were no longer the case, e.g., differences in access to training or conference funding. For some, this was a positive finding; for others, this was not the case. Onova, for example, noted that in moving institutions, she found little departmental support of postdocs, she was not allowed to teach because teaching assistantships were viewed as a source of funding for PhD students, and she could not access conference funding that both doctoral students and faculty could.

Policy Knowledge: Responsibilities and Resources

Policies at many levels and contexts influenced in positive and negative ways the kinds of work available to the postdocs, their status, benefits, salaries, and other gains as can be seen in these two cameos.

Earl, Social Scientist, 2 Years Postgraduation, Now in Europe

Earl got a 2-year postdoc contract immediately after graduation in his home country. Although he was expecting to do supervision, this kind of work was not recognized in his contract, so he chose to do only research. In the long term, he imagined a permanent position, but this might be abroad—likely in an English-speaking country, given the difficulties those in his country were experiencing.

Earl viewed policy as rules at national, institutional, departmental levels that are supposed to improve the quality of research and teaching. While he felt policy had not affected his career, he described an experience where a lack of policy led to inequity. He had applied for two academic posts as well as for the postdoc he got. He received news before the selection panel met that he had received the postdoc. The chair of the selection panel told him he was not chosen since he had the postdoc. He felt this was unfair—the committee didn't use objective criteria and consistent processes. "You can't trust the system"; this is one reason why he imagines going abroad.

As to influencing policy, he felt he needed power first—perhaps in 6–8 years he could influence the way in which selection panels work, and in 25 years as head of department he might make a number of things more transparent and equitable. In fact, he rarely thought or talked about HE policy.

Marian, Social Scientist, 2 Years Postgraduation, Now in Europe

Marian was unhappy that her postdoc contract had her teaching more than researching. Still, she had arranged a 2.5 month visit to North America, which she felt had enhanced her research profile. In the future, she hoped for her own funding supervising postgrads and collaborating internationally, but she would also be happy keeping her position, given the lack of contracts.

Marian believed policy should promote higher quality teaching and professional development of teachers and researchers—open up opportunities, not the reverse. She saw clearly how policies affected her work; for instance, the Bologna agreement required teachers to work together, but there was no institutional recognition or reward for this. Also, the requirements for tenure were not only extremely difficult for a newer researcher to meet, but the application was also opaque and seemed unfair since she has compared CVs of those who have been accepted and those not and could not see differences.

She also described how she had used policy (e.g., knowing about and applying for fellowships to finish her PhD) to go to England. She also noted a future possible HE policy that could have a big impact: a government move to have universities do more teaching and less research.

Cases developed from transcripts related to Ashwin, P., Deem, R., McAlpine, L.. 2015. Newer researchers in higher education: Policy actors or policy subjects? *Studies in Higher Education, 41*, 2184–2197.

Earl and Marian provide us with some sense of the experienced complexity of the relationship between work and policy, including the opaqueness of many of the policies. Generally, when individuals became (or were forced to be) aware of policies, most decided there was little they could change, even though the policies might be intrusive and debilitating. It was rare, indeed, to see individuals make strategic use of higher education policy, as Marian suggests she did.

In general, individuals were most aware of policies and procedures related to the research group, the department, and the local day-to-day working environment. For instance, they talked about the "politics" in their departments, in hiring processes and behind the scenes decisions—and the unfairness that occurred in these practices, as Earl and Marian report above. Reference was made to the lack of policies for ensuring the smooth running of the lab (Anne) and departmental policies, which affected employee status (Onova as a postdoc prevented from teaching). In other words, where individuals were most aware of policy impact was close to home, and recognition of the policy often seemed linked to a sense of exclusion (feeling a second-class citizen).

Institutional policies were also occasionally referenced (not infrequently linked to recognizing the difference in policy from a previous institution). Institutional policies that individuals referred to included the lack of criteria and guidelines for institutional career progression (Otto) and the failure to have policies for postdocs that offered the same benefits as for students and faculty (Jennifer). Very rarely did individuals note a positive policy when moving institutions. Victor was pleased that his new university had a policy of giving new people start-up funds so "I could immediately set up a small group and choose my own…directions to explore, and that was before I got the first external grant."

Reasons for this lack of knowledge about institutional policies might be explained by the fact that the postdocs were hired by a PI and tended only to interact with those in their research teams and sometimes department. They were not embedded institutionally, and other studies have suggested that postdocs often do not know of or take advantage of useful institutional policies, such as ones directed at offering training specifically for them (Scaffidi & Berman, 2011).

Postdocs' lack of awareness of regional or national policies was striking; it was only in relationship to research and fellowship funding where they appeared knowledgeable. For instance, Albert referred to changes in Canadian funding, which now permitted postdocs to be coinvestigators, whereas before they could not get credit even if they wrote the grant. Further evidence of the awareness of research funding policy emerged in the PI study. Many of the participants in this study knew about funding council policies related to supporting early career researchers, whether offered by the EU funding council or the funding councils of their own countries.[8] While the early career grants protected applicants from competing in the larger research pool, the criteria were designed to single out and support those perceived to have research leadership potential. As a result, lack of success could be perceived as a signal that they might not be competitive. Further, since the grants were only available for a limited period of time after PhD graduation and individuals could apply only a limited number of times, they had many factors to balance in deciding when was best to apply and how to focus the application.

---

8. Some were called grants, e.g., UK NIHR Career Development Fellowship, and EPSRC Leadership Grant, even though they might not include funding to hire and supervise others. Sometimes, there was a series of grants in which the expectations of research leadership expanded, e.g., in the Netherlands the Veni, Vidi, Vici program.

What should be apparent from this section of the chapter is the invisibility of the challenges to individuals and families in making mobility a priority for their careers. In other words, while it was perceived as advantageous (though providing no assurance of success), the reality is that for some, it might not have been possible; for others, it might bring a set of challenges they had not been prepared for. Further, as regards policy knowledge, it is of some concern how relatively ill-informed the postdocs were about the influence of policy on their work opportunities. Still, perhaps this is too much to expect since Brew, Boud, Lucas, and Crawford (2015) suggest that many midcareer academics in the United Kingdom and Australia may also focus little on policies and how they might influence investment in work. It would appear that policy and its influence on academic work may be a black hole—invisible to many and thus undiscussed despite its influence.

WHERE DO THEY FORESEE THEMSELVES IN 5 YEARS?

What does and can the future hold for those seeking tenure-track positions? In general, many continued to hope for such a position. Thus, there clearly was a mismatch between personal hopes and career opportunities, with many things beyond the individual's control influencing the career possibilities, especially given what Say referred to as "the infinite supply" of postdocs. In this section, I explore the experiences: first, of those who still retained hope after a number of years as postdocs and, next, of those who began to lose hope of a pretenure career and imagine a different kind of future.

Still Hope

Over the period of time we followed individuals in Study 1, each year we asked the following question: "where do you see yourself in 5 years?" Say's response each year was that "hopefully" he would have "secured an academic faculty position." Sustaining this hope was, indeed, necessary since in a few cases individuals waited 10 years after graduation before getting a tenure-track position. Reasons for this long wait could be linked to lack of mobility, research, and publication difficulties, or by the decrease over the past 2 decades in the availability of such positions. Fracatun realized that there were only four tenure-track positions globally that fit his research profile. Flora was strategic in seeking to change her research field to one that was growing and might, therefore, have more career possibilities. The downside was that her decision meant a longer time until she might be competitive enough for a tenure-track position. She recognized that she might still be a postdoc after 5 years unless her publishing went really well. As George said, "It depends a lot on what opportunities come up, but …most likely still doing research in some applied field, which I would be happy with." (He was not alone in seeking to be "happy" in whatever his future job might be.) What stands out is that the most agentive individuals incorporated a strategic view of career opportunities as well as worked to progress their own profiles toward their imagined futures.

Similarly, in Study 2, getting a grant as PI also took up to 11 years after graduation despite individuals being internationally mobile and applying many times. Otto and Romeo (first cameos) represent the average time to grant, with nearly half of the 60 participants obtaining a grant after 5 years. Notably, there was no statistical difference in time to grant with regard to (1) gender or (2) institutional location (where grant was awarded).

Why and how do they maintain their hope? My suppositions are the following. First, they loved doing research, so there were aspects of their work that were intrinsically motivating. Second, many had more senior colleagues in their networks who supported them, saw their potential, and sought opportunities for them; recall Romeo's story. Third, they had made a commitment to this career path, and changing career direction would be a psychologically if not socially challenging decision. And, fourth, individuals found a way to maintain motivation in the face of rejection, whether in relation to a grant, a publication or a job, by invoking luck.

Considering Alternatives

Most who were considering leaving academia had begun to view themselves as noncompetitive in the job market. Further, they had been postdocs for some time, had experienced a number of contracts, and generally were tired of dealing with the many challenges: the financial insecurity (AAA), concerns about families (Sandra), the lack of long-term benefits (Onova) and work–life balance (Charles). There was also a sense that they were being taken advantage of. For instance, Catherine became disenchanted over time given the repeated renewals of her contract over 10 years, and decided to go to the union to question why her position had not become permanent. Funky Monkey had begun to lose hope and said his future goal was to "either have a job in industry or a job in something related to…some sort of research…I don't mind doing stuff like maybe project coordinator or…project manager." At the same time, he resented the life he had given to science and that doctoral programs did not tell students about their real options.

How are these people different from those who were still committed to a tenure-track position after years as a postdoc? That is extremely difficult to assess since most exhibited the same personal commitment as those who continued to hope. However, there was often a negative aspect to their work environments (not often present in the stories of those who retained hope), which may have strongly influenced the decision to look elsewhere. Anne, for instance, regretted her choice of supervisor and lab; she felt ill-supported and dealt regularly with lab problems due to poor leadership. Charles found himself doing menial tasks, and Funky Monkey found his work worse than his PhD, which had been very difficult. Such experiences contributed to disillusionment about the nature and value of academic work. This combined with a growing feeling of not being competitive, which likely influenced the shift they made in imagining their future careers.

Unfortunately, however, most lacked knowledge of nonacademic careers. So, an openness to thinking beyond an academic position was not supported by information that would open up opportunities. Interestingly, there was little evidence that individuals undertook steps to research other possibilities; instead, they depended on anecdotal knowledge from their friends. Therefore, the decision to look elsewhere was driven by a loss of hope rather than one of moving to something new, different, and perhaps more interesting and appropriate for utilizing their gained experience. Others, the ones in countries seriously affected by the global economic crisis (i.e., budgets, positions, and salaries being cut regularly), were also dealing with the necessity of moving internationally to get an academic position, which they might not wish to do. This, again, made moving into the nonacademic sector a "push" rather than a "pull."

In thinking about the differences between the two groups, including how individuals managed their hopes for the future, Onova's case is striking. She had held three 1-year contracts where she benefited from increased collaborations, name recognition (supervisor and institution), and opportunities for papers in higher-status journals. Near the end of the third contract, she applied for and was invited to interview for three tenure-track positions, so she certainly appeared competitive. However, as her contract ended, she had had no news about the results, so she concluded she had not been successful. Having decided she did not want another postdoc (due to more moving, insecurity, etc.), she turned down the one position she was offered and started packing to return to her hometown and find some kind of work. It was just at this point that she was offered a tenure-track position that "exceeded my expectations" and in which she has been very successful. Her case is a reminder of the academic research potential that may be lost when individuals ultimately turn away from academia having decided they no longer want to wait.

Overall, the key structural problem these postdocs were facing was the pinch point of institutions being hesitant to make long-term funding commitments to individual careers while needing research labor to advance research productivity. In other words, multiple Master's students, PhD students, and postdocs are critical to this "industry," so there is a constant need for more of them. Yet universities want to limit their long-term risk. As a result, postdocs find themselves hostages to fortune.

IMPLICATIONS: WHAT MIGHT ALL THIS MEAN FOR THE FUTURE?

In this chapter, I set out to make some of the invisibility in postdocs' experiences visible, focusing on six aspects. We, whether researchers, faculty developers or policy makers, need to integrate and synthesize these various influences if we truly want to understand postdocs' work lives and career decisions. So, below I characterize the six themes and suggest how we might engage in change efforts.

First is the variation in the kinds of work postdocs do, which we (researchers, faculty developers and policy makers) often overlook. This is linked to a mismatch between sources of funding and a secure career trajectory since most postdocs are dependent on soft funding. Further, institutions may see little reason to invest, as postdocs are often viewed as passing through on the way to tenure-track positions elsewhere (Laursen, Thiry, & Loshbaugh, 2012). While this lack of career structure has been acknowledged in policy documents (for instance, the 2005 European Charter for Researchers and Code of Conduct for the Recruitment of Researchers), little has changed since the structure underlying research funding, and, thus, postdoc salaries, remains largely the same despite changes in fellowships that provide a staged approach to post-doc research. However, such programs are designed to winnow out the best. As well, it is clear that some universities (or at least departments within them) do try to provide contracts to sustain employment; in fact, we have examples where they did this, but it often meant that the postdoc involved would need to shift research direction or area. Ultimately the offers did not advance their careers. This issue is the most daunting since it is the most intractable. However, there are opportunities for institutional change in the following areas.

Second, agency and the role of emotion have not often been the focus of insti-tutional attention, but in my view, are central to any efforts to provide better sup-port to postdocs. If we accept the evidence that thinking and acting strategically are critical characteristics for researchers, and that part of being agentive is learning to cope effectively with such potentially damaging psychological reactions, we need to rethink the professional development offerings for both PhD students and postdocs. We need to examine our offerings by addressing this question: *How we explicitly help junior researchers be more strategic in developing their profiles as well as manage their emotions in useful ways to develop their emotional resilience?*

Third, the apparent lack of knowledge of alternate careers was an impedi-ment to individuals taking a more positive view of work beyond the academy (for doctoral students, see Jaeger, Haley, Ampaw, & Levin, 2013). *Doctoral programs and professional development for postdocs need explicitly to address nonacademic career options and academic career realities. Activities could include career panels where PhD graduates with different trajectories take questions; as well, formal opportunities and financial support for internships outside the academy could be offered given the evidence that such experiences may open up new career possibilities* (Cruz Castro & Sanz Menendez, 2005).

This career exploration and decision-making theme runs alongside another one, individuals' hopes and aspirations for their personal lives. A lack of align-ment between the two may lead individuals to turn away from an academic future. In other words, we cannot overlook the influence of personal factors in decision-making. *This suggests a need for greater awareness and integration into any institutional career support, whether policies or professional develop-ment offerings. It may be useful to review institutional human resource policies to ensure that postdocs are not disadvantaged as regards training and personal benefits. For instance, they could offer training specifically for postdocs and include offerings which incorporate dealing with family and personal issues, alternate careers, and counseling in and out of academia.*

Fifth, while mobility was perceived as valuable, it can be disruptive and is not always possible. If it is, indeed, as important as is believed (Kim & Locke, 2010), it may be necessary to rethink the nature of such opportunities. While most sojourners moved countries for lengthy periods of time, a few, like Marian and Romeo, did not and still found the experiences useful. *Creating funding opportunities for shorter international visits (with some preparation before-hand) might be one way in which such beneficial experiences could be more accessible and useful—both at PhD and postdoc levels.* It was interesting to note that those on fellowships often visited colleagues elsewhere for periods of not more than 1–2 months.

Sixth, there are implications related to policy knowledge and its influence on work. If more senior academics do not attend to policy and contest it, other than through silent resistance, there are two outcomes with significant impact on postdocs and other early career researchers. First, the conditions of work continue to change, often in a more constraining and time-pressured direction, creating worse work conditions and career opportunities for postdocs. Second, postdocs are neither explicitly learning about policy influences and changes nor how to contest them since they are not seeing this modeled. *Thus, there is a need to incorporate in both doctoral programs and postdoc professional develop-ment offerings that direct attention to the relationship between institutional and regional/national policies and academic work.*

APPENDIX A

TABLE 8.1 Postdocs in Longitudinal Study

Status, end of Study	No., Discipline Cluster	National	International	Male	Female	Recruited in Canada	Recruited in UK
Achieved research-teaching after postdoc	5 science	3	2	3	2	Onova, PhD, Sophia(e), Brookeye, Fracatun	
	4 social science	1	3	1	3		CM, Trudi, Paul, Jennifer
Not achieved research-teaching position but still hoping	9 science 7, 5 or <years 2, 6+years	7	2	7	2	George, Flora, Tina, Say, AAA Thor Bear, Albert	TDB, SA
	3 social science 2, 5 or <years 1, 6+years	2	1		3		KS, Elizabeth Catherine
Turning away from research-teaching after postdoc	1 science	1		1		Funky Monkey	
	2 social science	2		1	1	Charles, Ann	
Total	24	16	8	13	11		

TABLE 8.2 Demographic Characteristics by University

University	Total/Gender	Origin	Disciplinary Cluster: Sciences, Technology, Engineering, Math and Medicine (STEMM); Arts, Humanities and Social Sciences (AHSS)	Years to Grant From PhD Graduation: Average, Variation
UK 1	19/13M, 6F	7 national, 12 international	10 STEMM, 9 AHSS	6.22 years, 0–11 years
UK 2	25/15M, 10F	13 national, 20 international	16 STEMM, 9F	4.61 years, 1–10
EU	18/12M, 6F	13 national, 5 international	11 STEMM, 7 AHSS	4.58 years, 1–9 years
Total	60/39M, 21F	33 national, 27 international	36 STEMM, 24 AHSS	5.08 years, 0–11 years

TABLE 8.3 Postdoc Characteristics (#, %)

Gender	Male	Female			
Survey, 65 out of 71 (missing values 6)	23, 35%	42, 65%			
Interviewees	3, 30%	7, 70%			
Disciplinary Cluster	**Social Sciences**	**Science, Technology, Engineering, Math, Medicine**			
Survey	26, 36.62%	45, 63.38%			
Interviewees	7, 70%	3, 30%			
Years of Postdoc Experience	**<1 year**	**1–2 years**	**3–4 years**	**5–6 years**	**>7 years**
Survey	1.6%	24.2%	29%	12.9%	32.3%
Interviewees	1, 10%	5, 50%	1, 10%	2, 20%	1, 10%
Origins	**The United Kingdom**	**Europe**	**North America**	**Other**	
Survey 65 (missing values 6)	25, 38.4%	26, 40%	5, 7.6%	9, 13.8%	
Interviewees	2, 20%	6, 60%	1, 10%	1, 10%	

TABLE 8.4 Characteristics of Higher Education Researchers

Characteristic	#, %
Location	
Europe	13, 31%
UK and Ireland	21, 50%
Others globally	8, 19%
Role	
PhD	29, 83%
Postdoc	13, 13%
Gender	
Female	31, 74%
Male	11, 26%

REFERENCES

Archer, M. (2003). *Structure, agency and the internal conversation*. Cambridge: Cambridge University Press.

Ashwin, P., Deem, R., & McAlpine, L. (2015). Newer researchers in higher education: Policy actors or policy subjects? *Studies in Higher Education, 41*, 2184–2197.

Bagnoli, A. (2009). Beyond the standard interview: The use of graphic elicitation and arts-based methods. *Qualitative Research, Special Issue on Qualitative Research and Methodological Innovation, 9*, 547–570.

Baruch, Y. (2004). Transforming careers: From linear to multidirectional career paths. *Career Development International, 9*, 58–73.

Bazeley, L. (1999). Continuing research by PhD graduates. *Higher Education Quarterly, 53*(4), 333–352.

Brew, A., Boud, D., Lucas, L., & Crawford, K. (2015). Responding to university learning teaching and research policies and initiatives: The role of reflexivity in the mid-career academic. *European Association for Research Into Learning and Instruction* (Larnaca, Cyprus).

Cantwell, B. (2011). Academic in-sourcing: International postdoctoral employment and new modes of academic production. *Journal of Higher Education Policy and Management, 33*, 101–114.

Cote, J. (2005). Identity studies: How close are we to developing a social science of identity? – An appraisal of the field. *Identity: An International Journal of Theory and Research, 6*(1), 3–25.

Cruz Castro, L., & Sanz Menendez, L. (2005). The employment of PhDs in firms: Trajectories, mobility and innovation. *Research Evaluation, 14*, 57–69.

Elliott, J. (2005). *Using narrative in social research: Qualitative and quantitative approaches*. London, UK: Sage.

Jaeger, A., Haley, K., Ampaw, & Levin, J. (2013). Understanding the career choice for underrepresented minority doctoral students in science and engineering. *Journal for Women & Minorities in Science & Engineering, 19*(1), 1–16.

Kim, T., & Locke, W. (2010). *Transnational academic mobility and the academic profession.* London: Centre for Higher Education Research and Information, Open University.

Laursen, S., Thiry, H., & Loshbaugh, H. (2012). *Mind the gap: The mismatch between career decision-making needs and opportunities for science of PhD students.* Vancouver, Canada: Paper presented at the American Educational Research Association.

McAlpine, L. (2010). Fixed-term researchers in the social sciences: Passionate investment yet marginalizing experiences. *International Journal of Academic Development, 15*, 229–240.

McAlpine, L. (2012). Academic work and careers: Re-location, re-location. *Higher Education Quarterly, 66*, 174–188.

McAlpine, L. (2016a). Becoming a PI: Shifting from 'doing' to 'managing' research. *Teaching in Higher Education, 21*, 49–63.

McAlpine, L. (2016b). Why might you use narrative methodology? A story about narrative. *Eesti Haridusteaduste Ajakiri. Estonian Journal of Education, 4*, 32–57.

McAlpine, L. (2016c). Navigating careers: Perceptions of post-PhD researchers. In *Paper presented at the higher education conference, Amsterdam, Netherlands.*

McAlpine, L., & Amundsen, C. (2016). *Post-phd career trajectories: Intentions, decision-making and life aspirations.* Basingstoke, UK: Palgrave Pivot.

McAlpine, L., Amundsen, C., & Turner, G. (2014). Identity-trajectory: Reframing early career academic experience. *British Educational Research Journal, 40*, 952–969.

McAlpine, L., & Mitra, M. (2015). Becoming a scientist: PhD workplaces and other sites of learning. *International Journal of Doctoral Studies, 10*, 111–128.

McAlpine, L., & Norton, J. (2006). Reframing our approach to doctoral programs: A learning perspective. *Higher Education Research and Development, 25*(1), 3–17.

McAlpine, L., Turner, G., Saunders, S., & Wilson, N. (2016). Becoming a PI: Agency, persistence, and some luck!. *International Journal for Researcher Development, 7*, 106–122.

Pedersen, H. (2014). New doctoral graduates in the knowledge economy: Trends and key issues. *Journal of Higher Education Policy and Management, 36*, 632–645.

Powell, K. (2015). The future of the postdoc. *Nature, 520*(7546), 144–147.

Scaffidi, A., & Berman, J. (2011). A positive postdoctoral experience is related to quality supervision and career mentoring, collaborations, networking and a nurturing research environment. *Higher Education, 62*, 685–698.

Skakni, I., & McAlpine, L., (in press). Post-PhD researchers' experiences: an emotionally rocky road. *Studies in Graduate and Postdoctoral Education.*

Sondergaard, D. (2005). Academic desire trajectories: Retooling the concepts of subject, desire and biography. *European Journal of Women's Studies, 12*(3), 297–313.

Van Arensburgen, P., & Van den Besselaar, P. (2012). The selection of scientific talent in the allocation of research grants. *Higher Education Policy, 25*, 381–405.

Weick, K., Sutcliffe, K., & Obstfeld, D. (2005). Organizing and the process of sensemaking. *Organization Science, 16*, 409–421.

Chapter 9

Global Perspectives on the Postdoctoral Scholar Experience

Karri Holley[1], Aliya Kuzhabekova[2], Nick Osbaldiston[3], Fabian Cannizzo[4], Christian Mauri[5], Shan Simmonds[6], Christine Teelken[7], Inge van der Weijden[8]

[1]University of Alabama, Tuscaloosa, AL, United States; [2]Nazarbayev University, Astana, Kazakhstan; [3]James Cook University, Cairns, QLD, Australia; [4]Monash University, Melbourne, VIC, Australia; [5]Murdoch University, Perth, WA, Australia; [6]North-West University, South Africa; [7]V U University Amsterdam, Amsterdam, the Netherlands; [8]Leiden University, Leiden, the Netherlands

Chapter Outline

Australia	204	Reflections on the International
Kazakhstan	208	Experiences of Postdoctoral
The Netherlands	213	Scholars 221
South Africa	217	References 223

The postdoctoral scholar exists globally within rapidly changing institutional contexts. While the role of postdoctoral scholars varies according to the national system, shared concerns exist, including uncertain career and employment opportunities as well as the neglect of professional development (Åkerlind, 2005). These concerns are magnified by the challenge of defining the postdoctoral role. One definition might assume that a postdoc is a research-only position to be held by a PhD recipient, while another might emphasize the early career nature of the position as a stepping stone toward a more permanent faculty role. The nature of academic work cannot be understood without documenting local, national, and international changes such as the perception of higher education as a public good, the rate of economic expansion, the desire for institutional prestige, globalization, and massification (Cummings & Teichler, 2015).

This chapter examines the postdoctoral scholar experience in four countries: Australia, Kazakhstan, the Netherlands, and South Africa. Each country-specific essay below introduces the unique national system and identifies challenges facing postdoctoral scholars within the system. The respective higher education

The Postdoc Landscape. http://dx.doi.org/10.1016/B978-0-12-813169-5.00009-4

systems exhibit different pathways of development as well as unique priorities related to teaching and research for domestic and international students.

Across the global stage, academic institutions exist within as part of a center/periphery dichotomy. Altbach (2016) suggests that "central" academic institutions are research-oriented with high levels of prestige and membership in the international knowledge system, while "peripheral" academic institutions serve to distribute knowledge through the training of students. Academic institutions at the center dominate global rankings, while those at the periphery tend to be located in emergent or developing economies. Boundaries between the two institutional types are blurred by isomorphic efforts to accrue organizational prestige and intellectual resources, increasing the tendency of institutions to mirror structural and organizational forms. The postdoctoral scholar signifies as one of these forms. Of interest is how national systems with various degrees of maturity exhibit shared characteristics related to defining, supporting, and promoting the postdoctoral role. These shared characteristics emphasize the nature of colleges and universities as global institutions.

AUSTRALIA

The Australian higher education system is driven today by a neoliberal agenda that guides early career researchers toward highly individualized approaches to their careers. Three facets underpin the postdoctoral experience and produce significant strain on individuals. First, the lack of job prospects provokes a sense of competition among early career researchers wherein expected levels of accomplishment for entry into academia have vastly increased over the past decade. Second, the ideals of academia and the nostalgia for times past provoke a sense that academia is no longer as solid as it once was. Early career researchers/postdoctoral researchers in long-term (or even ongoing permanent) positions feel that institutions could alter the arrangement at any time. Last, career development ideals place strain on early career researchers' experiences of time. The increased pressure to spend time on activities that are important for institutional needs (such as teaching or administration) has meant that time spent on substantive or authenticating activities such as research is limited.

Australia's higher education landscape is similar to other western modernized countries such as the United Kingdom. In the past, the policies of the State have been oriented to a supply-driven model wherein funding was provided to institutions individually according to defined needs in research and education provision. In 1988, the federal government reformed the *Higher Education Funding Act*, colloquially known as the "Dawkins reforms" after the minister who led the changes. The reforms were initiated in an attempt to "support a higher education system that" would be "characteristic of quality, diversity and equity of access" and further "strengthen Australia's knowledge base and enhance the contribution of Australia's research capabilities" (Higher Education Funding Act, 1988). The reforms fundamentally changed the Australian system by turning it into a demand-driven model wherein funding particularly hinged on student enrollments. At both a teaching and research level, this model exposed the universities to a new level of competition. Broader workplace legislation reforms, which promoted a casualization of the workforce across a number of sectors and withdrew the long-term stability once associated with vocations such as teaching and education, correlated significantly with these higher education changes (Coates & Goedegebuure, 2012; Gottschalk & McEachern, 2010; Kenway, Bullen, & Robb, 2004; Marginson, 2008; NTEU, 2012; Ward, 2012).

The turn is best described as a shift toward neoliberalism (Cannizzo, 2017; Davies & Bansel, 2005, pp. 47–58; Marginson, 2008). For the university, this means that students become customers, measured in terms of income to the institution. Subsequently, courses become products for marketing within the society where they are located. In Australia, this competition among higher education providers is significant with 43 accredited universities[1] all vying for potential students using different marketing techniques. These sorts of structural pressures impact individual academics within and outside of the system; a neoliberal university, or, indeed, workplace, becomes exposed to both competition for resources and further accountable for how these resources are used within an audit culture. Schools, colleges, and faculties are measured by their income to the university (predominantly through how much wealth they create through teaching), which then exposes courses and subjects to scrutiny for their viability within the market. Internally, this pressure constructs powerful contestations within the university for resources, including limited pools of funds available for research that arguably provide entry-level positions for postdoctoral researchers. All these structural changes have made the individual academic more visible in terms of measuring, calculation, and accountability (Cannizzo, 2017; Davies & Bansel, 2005, pp. 47–58; McWilliam, 2004). Outputs, such as research publications, grants, and teaching statistics (especially student survey data on teaching performance), become markers of individual success and are used to compare one academic with others (Kelly & Burrows, 2012).

Younger generations coming through the PhD process develop and construct their careers in a far different environment structurally than previous generations. In Cannizzo's (2017, p. 14) study, early career academics in Australia constructed their careers with emphasis on passion and authenticity, but they were far more likely to "account for their career narratives in terms of survival." Subsequently, this group is then tuned into the neoliberal practices in a way that "normalises compliance with managerial imperatives as an unavoidable externality" (Cannizzo, 2017, p. 14). While a certain understanding of the

---

1. This includes mostly State-run universities that are publically funded alongside a relatively small number of international/private-run institutions. In addition to this, there are several small privately owned and operated training organizations and State-run technical institutions (TAFES) that teach workplace-based diplomas. However, funding for these smaller institutions has begun to be withdrawn from the sector with a greater emphasis on private providers and competition.

nature of neoliberalism and its problems among postdoctoral researchers exists, an implicit acceptance of it also exists within career narratives, wherein these practices become part of the making and marking of progress (Archer, 2008; Cannizzo, 2017). Early career researchers become an active part of the system and thus constitute the neoliberal university.

Neoliberal strategies of governance have raised the bar considerably for postdoctoral researchers seeking employment in academia. Selection criteria for vacancies are now replete with questions related to performance measures. Publication points, grant income, teaching viability (proven through student statistics and evaluations), and the potential to accrue more of these for the department serve as the components of battleground on which postdoctoral researchers compete for jobs with one another. Although anecdotal at times, there is a sense within the current research that as generations of PhD students graduate, the standard for entry into a position within academia increases. Institutional pressures to ensure that any potential employees are aligned with strategic goals of the universities and have demonstrated excellence in research and teaching (as well as the potential to build on this excellence) mean that PhD students have very limited option but to buy into the neoliberal agenda.

PhD graduates, therefore, end up entering into an academic job market that is highly contingent on resource funding (especially in the sciences), increasingly leaning toward short-term contract or casual labor, and far more competitive than years past. Furthermore, the ratio of graduates to job vacancies in academia has grown considerably. As Crossley (2013) demonstrates, the number of PhD graduates in Australia is far greater than in the past. The shift toward more of a demand-driven model for education has accelerated the amount of people attending and graduating with a degree and also increasing those undertaking and completing a PhD. Australian graduates are also competing with a globally mobile workforce, with some universities in Australia valorizing their international presence and, therefore, actively seeking overseas applicants. Subsequently, PhD students need to be able to have skills to find employment outside of the academic environment—the jobs are not necessarily available within academia anymore. In a sense, younger academics are now becoming more flexible with their qualifications, and new models of completing the PhD have accentuated skills that align with a number of careers (not just academia). Yet this sort of flexibility is indicative of the broader trend in the neoliberal economy.

The casualization of the workforce in academia is also an increasing issue facing postdoctoral researchers finding their way career-wise. Australian universities are becoming heavily dominated by cheaper short-term labor, especially in relation to teaching. Casual academics now account for 22 percent of the academic workforce in Australia (May, Strachan, & Peetz, 2013). This number is likely to mask the *real* figure, however, as the data are based on full-time equivalent calculations that "equate the hours of a casual academic with that of a full time academic" (May et al., 2013, p. 3). May and colleagues find in their quantitative work that the figure is probably more likely greater than 50%. This figure is significantly larger than the broader Australian labor figures that places casual labor at around 24% of the total labor force. Unfortunately, early career researchers likely find themselves within sessional/casual academic markets with one survey conducted by the national union concluding that approximately 29% of their respondents were PhD holders and were actively pursuing an academic career (NTEU, 2012). As Natalier et al. (2016, p. 3) argue, postdoctoral researchers in this space are "often not recognised as members of the institution and disciplinary communities." Subsequently, access to resources such as funding for networking (for instance, conferences) is not available for the casual employee, which influences career progression. Once embedded into the sessional market, the early career researcher can, as Natalier et al. (2016) warn, end up working for some time before progressing inside the academy. The sessional labor is heavily dominated by women (NTEU, 2012). Gender barriers to employment in higher education are made more problematic when care duties such as child rearing affect women's ability to enter the market. Institutions still have limited options for women looking to come back from maternity breaks, specifically those who left during or after their postgraduate research.

Even if the early career researcher is able to secure employment within the university, there is a growing sense that due to the processes mentioned earlier, uncertainty remains a part of the academic life. Our research suggests that postdoctoral researchers and teachers see the academic career as ill-defined compared to the past (Cannizzo, 2017; Osbaldiston, Cannizzo, & Mauri, 2017). Early career researchers seem well aware of how things are and the uncertainty in the present sector. In particular, the closure of faculties, departments, courses, and even subjects provokes feelings of uneasiness. The manner in which individuals are exposed through metrics and the calculative nature of the academy now mean that any notion of permanency can be cut quickly. In Australia, recently, a number of universities have, indeed, held redundancy rounds wherein staff who have been well established found themselves unemployed. As we have found in our research, younger academics are not unaware of these issues, and when older mentors, in particular, are victims of these trends, they see the academic career as one that requires flexibility and reflexivity. They understand that to remain employable, they have to make their research visible and keep their performance indicators in line with not only what the university requires of them but also to sustain their research impact at a national/international level to remain competitive. Cannizzo (2017) finds in his research that early career researchers are far more likely to narrate their careers in terms of survival.

Once inside that system, an ongoing conflict exists between how time is split for the early career academic. Postdoctoral researchers, in particular, face greater pressure than their predecessors did to spend their time in formal/instrumental labor such as teaching, administration, and service within their institutions. The ability to control these "times" is difficult for academics generally in the neoliberal university, as there is an "increasing tempo and intensification" of these activities (Cannizzo & Osbaldiston, 2016, pp. 890–906; Osbaldiston et al., 2017, p.13). Younger academics tend to see time that is devoted to research and writing as the more authenticating than those of teaching, administration, and service. Rather, instrumental labor is occasionally suggested as those items that need to be done quickly or managed as effectively as possible to enable more of the research time that is inherently fulfilling.

Early career researchers are also forward thinkers with ideals about where they see themselves in the future determining what they value in the present (Osbaldiston et al., 2017). They are savvy and understand that the neoliberal institution counts certain activities such as research publications and grants and thus places high value on these in measuring individual academics. Specifically, early career researchers in our studies see their academic career success through things such as employment opportunities elsewhere and internal promotions as hinged on research success (such as books, grants, and papers). While time spent in researching and writing promotes a certain intellectual authenticity within the individual, the same time aligns well with the neoliberal mechanisms of the university. Subsequently, as Archer (2008) and Cannizzo (2017) argue, the early career academic has, in large part, become an active player in the neo-liberal transformation of the Australian university system.

The situation in Australia is complicated even further now by the call for a total deregulation of the higher education sector, which would allow universities to set their own prices and, in some cases, drastically increase costs for students in the long term. Some of our participants in studies conducted recently have suggested that these sorts of measures make them very cautious about the future. Early career academics in particular are worried about the implications of making the system even more demand driven and customer focused. They face multiple pressures just to enter into and then sustain an academic career. However, the changing system means they are more uncertain than their predecessors about the future of academia and whether staying inside of it will be sustainable.

KAZAKHSTAN

After gaining independence as a result of dissolution of the Soviet Union in 1991, the Central Asian Republic of Kazakhstan implemented a variety of reforms to adjust its educational system to the requirements of the new political and economic reality. Much of the national reform agenda was motivated by the desire to transition from a centrally planned economy serving the needs of the industrial enterprises in the Soviet center to a market-driven knowledge and innovation-fueled economy following the example of "Asian tigers" such as Japan, Hong Kong, Taiwan, and Singapore (Aitzhanova, Katsu, Linn, & Yezhov, 2014). In higher education, modernization efforts were largely determined by the standards adopted through the Bologna Process, which Kazakhstan officially joined in 2010 (Tampayeva, 2015). Higher education underwent privatization with the emergence of the private sector and introduction of student fees; internationalization via the introduction of mobility programs as well as the three-level degree system (Bachelors, Masters, PhD) to facilitate degree acceptance and credit transferability to the European Union; transition to independent accreditation as the main mechanism for quality control; introduction of standardized entrance exams and standards-based educational curriculum; and development of greater university autonomy and increasing faculty control over the curriculum and academic matters.

While modifying the methods of instruction and curriculum content at the undergraduate and master's level has been somewhat successful with universities producing professionals demanded by the market (instead of preordered by a command economy), the preparation of PhD-trained researchers and the research capacity of postindependence and Soviet-trained faculty continue to face numerous challenges. Such challenges include poor access to research facilities, equipment, and library resources at universities as well as the poor availability of project funding (World Bank, 2007). Many of the challenges are a direct result of the Soviet legacy, whereby universities and university faculty were primary responsible for teaching and training of the nonscientific workforce. University faculty existed, with the notable exception of leading national research universities, in relative isolation from the research and innovation system (Smolentseva, 2003). The latter system was composed of (1) the industrial enterprise research labs, engaged mostly in applied research and development, and (2) a system of research institutes within the Academy of Sciences, which was primarily responsible for basic research, dissertation supervision, and apprenticeship training of researchers. Most Soviet universities engaged predominantly in basic research and lacked the infrastructure and facilities of non-university research centers. These research centers were more popular among trainees seeking PhD degrees.

Organized postdoctoral training did not exist in the Soviet Union for two reasons. First, the Soviet doctoral degree was different from the PhD degree as conceptualized in Western countries. The degree was awarded to very few individuals; the dissertation was written independently over the course of many years and without specialized formal coursework or training. The terminal degree was awarded to seasoned accomplished researchers who already had formal researcher positions and who did not require any additional (postdoctoral) training. The Western-style PhD degree was more similar to Candidate of Science degree, for which a person was required to complete coursework and to defend a supervised dissertation. Second, even after obtaining the Candidate of Science degree, a person did not complete anything similar to a Western postdoctoral training program because the academic profession did not require a human capital enhancing transition stage between degree completion and the job search in a competitive market. In the Soviet Union, the labor market in its conventional competitive form did not exist. Faculty were often "inbred" or grown from within by gradually involving internal Candidate of Science students into teaching. Researchers in research institutes were selected to fill available positions from a pool of internally trained students or students from a network of collaborators according to a centrally determined hiring plan by the government after completing a Candidate of Science degree. Available positions in research institutes comprised a hierarchy (junior researcher, researcher, senior researcher, lead researcher, laboratory head), and promotion was determined by the years of research experience and successful defense of a doctoral dissertation.

The separation of teaching and research functions continues to persist in contemporary Kazakhstan. While the Kazakhstani Ministry of Education seems aware of the importance of integrating teaching and research, and of developing the faculty's research capacity, in most universities, faculty remain relatively uninvolved in research (World Bank, 2007). Despite the fact that the postgraduate degree structure has changed as a result of the Bologna Process implementation and the Soviet-style Candidate of Science and Doctor of Science degrees have been mostly replaced by a Western-style PhD degree, postdoctoral training has not been adopted by Kazakhstani universities with the new degree structure. As in the Soviet Union, a post-PhD degree research apprenticeship is not expected from future faculty since teaching remains the main responsibility of faculty and most faculty are in-bred. The apprenticeship is also not expected to be hired at research institutes, where the conventional hierarchy of research positions and the system of internal training continue to exist.

A postdoctoral degree has been gradually introduced at Nazarbayev University (NU), which was established in 2011 to become the first world-class research university in the country. This early experience with postdoctoral positions and the conditions of work of postdoctoral students at the university present a case for analysis. Lessons from the experience of NU may point to any potential issues and challenges as well as possible solutions that might inform other post-Soviet universities (or those in non-Western contexts) considering the introduction of postdoctoral positions.

Created in 2011 through the initiative of Kazakhstani President Nursultan Nazarbayev, NU was established in collaboration with reputable research-intensive partner universities from around the world (such as the universities of Cambridge, Pennsylvania, Duke, etc.) (OECD, 2017). The partners actively participate in setting up academic processes and curricula, quality review and monitoring, collaborative research projects, hiring, and the professional development of academic staff. The university uses a Western-style degree structure and curriculum, which are delivered by the predominantly internationally hired faculty (Jumakulov & Ashirbekov, 2016). The majority of students at the university come from domestic middle- or upper-middle class families; they have previously experienced a very good quality of education; and, in general, they have higher grades and better school exit and university entrance scores compared to students in other universities. The university enjoys autonomy from the Ministry of Education and receives more generous research funding from the government of Kazakhstan compared with typical higher education institutions (Law on Status of Nazarbayev University Nazarbayev Intellectual Schools and Nazarbayev Fund, 2011).

Unlike faculty at domestic universities, NU faculty are expected to be both active teachers and active researchers. The majority of faculty have a Western PhD degree and are hired from international staff with very few exceptions. Faculty have a high status at the NU and receive an internationally competitive salary and benefits package, which is extremely generous by Kazakhstani standards. In addition to academic departments, NU has a system of research institutes and research centers, which are only formally affiliated with academic institutions and hire full-time research staff. These staff do not engage in teaching and may have either Soviet Candidate of Science or Doctor of Science degrees (for older generations) or post-Soviet PhD degrees earned domestically or abroad (for younger generations). These researchers are hired mostly from among Kazakhstani citizens. They are promoted based on the years of experience as researchers and their level of academic degree, following the tradition for the post-Soviet hierarchy at research institutes. Finally, while high by domestic standards, their salary remains much lower compared with the salary of the internationally hired faculty.

The first postdoctoral positions at NU were not necessarily called as such because they were introduced on an ad hoc basis. The responsibilities included (1) supervision of more junior researchers (typically graduate students); (2) design, planning, and implementation of a research project or a project funding application with a relatively small amount of guidance and supervision from their project leader; (3) performance of some teaching responsibilities under supervision of a more senior faculty; and (4) participation and presentation of project results at academic conferences. Their individual goal for being hired as a postdoc included gaining additional research experience as an apprentice of an established researcher before assuming a position as a faculty or more senior-level researcher at the university.

Given the fact that most schools at NU have only started to offer PhD programs and do not have a good supply of research assistants among PhD students, the main reason for faculty and lead researchers to hire postdocs was to staff their funded research projects with a skilled workforce of people who can perform advanced level research tasks. Postdocs with Western PhD degrees were hired in some cases with an explicit goal to facilitate the development of the research capacity of local junior researchers without Western PhD degrees by providing project oversight, guidance, and direct skill development training. In other cases, postdoctoral positions were used to enhance the research capacity of local researchers with Western PhD degrees who did not yet qualify for faculty positions due to their lack of experience and publications. In still other cases, postdocs were hired from among local domestic and international PhD degree holders to develop into a future full-time researcher within the nonacademic system of research centers and institutes.

When the human resource structure and classification of hiring positions were conceptualized in the early years of the university, the postdoctoral position was not included as a full-time position. Hence, no permanent contract arrangement exists to hire postdocs; renewable short-term contracts are the only option. According to Kazakhstani immigration law, foreigners can be issued a visa only if they are hired on long-term contracts. Exceptions are made only for those foreigners who are hired as international experts (for whom special types of short-term contracts are available). Under this circumstance, most of the postdoctoral positions could be filled only with local candidates. In some cases, international postdocs were hired on short-term contracts when they did not need a visa to enter Kazakhstan. However, according to the Labor Code, those on short-term contracts could not be offered a benefits package. They also faced the inconvenience of being paid in several large versus regular monthly installments and going through the hassle created by the need to file complicated monthly performance reports and the necessity to resign contracts. Some of the international postdocs hired on the short-term contracts, after a period of "probation," were moved to long-term positions as faculty. Others chose to leave due to complications created by the contractual arrangement and the associated immigration regulations.

Meanwhile, hiring local researchers as postdocs also presented challenges. Very few Kazakhstani students choose to pursue a PhD degree in Western countries. Even when a project leader manages to find a local PhD holder with the necessary set of skills, the leader faces difficulties in attracting him or her to an unstable short-term position available at NU. Some project leaders, representing primarily sciences, may find local Western PhD degree holders not only unprepared to be faculty but also underqualified to be postdocs.

One of the explanations for this challenge emerged through recent interviews with faculty at NU. Most local Western PhD degree holders were financially supported by the presidential scholarship "Bolashak" (Future) for the duration of their study with an obligation to return and work in Kazakhstan for 5 years. Under this plan, Bolashak scholars pursuing a PhD did not always hold research assistantships since they already had financial support. Because of the obligation to return to Kazakhstan, the Bolashak scholars also had low motivation to actively publish and develop their capacities to become competitive as future faculty in the international academic market. For many Bolashak scholars, the PhD was the first degree pursued abroad, so during the initial years of their studies, they focused more on improving their English language proficiency as opposed to gaining research experience. Hence, as our interviewees concluded, the Bolashak students had weaker skills and experience than their Western counterparts.

Most of the local postdocs are hired on short-term contracts without benefits within the system of research centers and institutes, and they are funded from grants of the Kazakhstani Ministry of Education and Science. In Kazakhstan, the Ministry of Education grants are reconsidered every year of the project; funding may be discontinued in the middle of the project despite the fact that a postdoc was hired for the whole duration of the study. The Ministry of Education and Sciences funding scheme also follows the national budget cycle, which involves a long unpaid wait period between the Ministry's decision to fund a project and the date when the awarded amount is allocated to the recipient's account. The postdoc scholar must find monies during the lengthy wait period when they are not paid a salary. Local postdocs have two types of career aspirations. Those who do not have a Western PhD tend to be interested in continuing their career as a full-time researcher. Their chances of being hired as a faculty are very small, so they aspire to expand their publication record and experience to receive a highly desirable position at NU, at least within the hierarchy of a research center. Western PhD holders typically aspire to become NU faculty and, at times, consciously try to supplement their postdoc experience with some part-time teaching.

In summary, the early experience with postdoctoral positions at NU was associated with several problems. Some problems are related to constraints of the local labor market and the immigration barriers preventing hiring postdocs from abroad. Other problems are associated with a lack of contractual and financial arrangements to ensure sustainability and security of postdocs hired to work on university projects. At the individual level, the absence of proper contractual and financial arrangements stripped away any sense of job security and made researchers feel unsafe about their future employment prospects.

As NU started to place greater emphasis on its research mission in recent years, a university-level discussion evolved about the necessity of introducing full-time long-term postdoctoral researcher positions. A policy regulating employment conditions for postdocs is being currently developed by a task force group created by the Faculty Senate. This development implies that the university is learning from its past mistakes and that the postdoctoral position should be a permanent addition to the Western-style research university. However, under the current structure of the Kazakhstani higher education system and the separation between the teaching and research function, the extent to which postdoctoral position may be adopted by other universities is not clear.

THE NETHERLANDS

The Dutch system of higher education involves a binary structure: 37 universities for applied sciences (HBOs) and 14 research-oriented universities, which have the exclusive right to confer doctorate degrees. In terms of performance, the Dutch academic landscape is often described as a high plateau with a few peaks (Adviesraad voor Wetenschaps-en Technologiebeleid, 2014). As in other European countries, the higher education system in the Netherlands is increasingly seen as being influenced by private sector elements (Boyne, 2002). This implementation of New Public Management (NPM) should lead to a more market-oriented higher education system, one that is able to compete for clients, funding, and prestige and to meet the growing pressure to cut costs

(Christensen & Lægreid, 2001; Scharitzer & Korunka, 2000). A key feature of NPM is its focus on performance in all aspects of management, primarily through NPM instruments such as pay for performance, performance appraisal, performance budgeting, and performance indicators. Efforts to develop new modes of academic governance have begun to significantly affect academic work and employment relations (Kehm & Lanzendorf, 2006).

The focus on performance in the Dutch higher education system resulted in an increasingly output-oriented system for the financing of academic research. This focus encouraged an increased number of PhD trajectories, as these are, for a large part, financed on output; in other words, universities obtain funding on the basis of their numbers of awarded PhD titles. This focus also resulted in more and longer postdoc trajectories as the financing of research is increasingly project based, leading to more temporary contracts.

The standard upward career trajectory in Dutch academia supposedly depends on a combination of both the individual merits and the positions available in the science system (Waaijer, 2016) with five career steps: full professor; associate professor; assistant professor; other scientific staff including postdocs; and the PhD candidate. The number of PhD graduates from Dutch universities has been increasing over the past 10 years (De Goede, Belder, & de Jonge, 2013); 6.6 of every 1000 economically active Dutch residents has a PhD, which is slightly lower than the EU-15 average of 7.5. The number of doctorate conferrals has doubled since 2000, which will consequently improve the relative position of the Netherlands. The share of women among doctorate recipients has also been increasing, from less than a quarter in 1990 to almost half in 2010 (De Goede et al., 2013). Furthermore, internationalization has taken place; whereas in 2003, 64 percent of PhD candidates were of Dutch nationality, this percentage dropped to 57 by 2011 (De Goede et al., 2013). These developments mirror international trends regarding the achievement of gender balance (Auriol, Misu, & Freeman, 2013) and an increasing importance of foreign-born workers in the academic workforce (e.g., Stephan, 2012 for the United States).

Opportunities for academic employment have not kept pace with this increase in the number of PhD graduates in the Dutch higher education system (Waaijer, 2016). Over the years, competition for positions in the Dutch academy has intensified. A postdoc function in the Netherlands is no longer a short-term entrance trajectory into the university but is more likely a longer-lasting experience, as postdocs tend to work on a series of temporary contracts, which becomes a preliminary requirement for obtaining a permanent position in the academy (Van der Weijden, Belder, Van Arensbergen, & Van den Besselaar, 2015). As a result, the postdoc population in the Netherlands is substantial and extensively growing. In 2005, 2559 postdocs existed; this number increased to 3463 in 2011 (Van Arensbergen, Hessels, & van der Meulen, 2013). As these numbers exclude medical sciences, we estimate a larger number of postdoctoral researchers in the Netherlands. The duration of their postdoc employment is approaching the length of the PhD trajectory (=48 months; Van der Weijden et al., 2015).

Their career prospects within the academy are weak. Annually, 750 assistant professorships are available, covering about 20% of the postdoc population in the Netherlands (De Goede et al., 2013).

Several challenges face Dutch postdoctoral scholars (Van der Weijden et al., 2015). The first challenge is the position of postdocs within their organization combined with the limited knowledge on postdoc demographics. Influenced by the sharply rising number of postdocs, universities have not been able to keep up with basic administration, causing a lack of knowledge on their demographics (Davis, 2009). Further, postdocs do not have their own category in the official formation system, which makes them harder to monitor. Postdocs seem to fall in a "gap" between being a (PhD) student and an employee, therefore not reaping the benefits of being part of either group (Davis, 2009). A thorough scan of the websites of four Dutch universities showed that postdocs are considered as temporary or nontenured staff, and several forms of career guidance, mentoring, and specific courses are mentioned; in many cases, PhD students and postdocs are considered as one group of employees (Ruiter, 2015).

Although professional managers and staff comprise the contemporary Dutch university, Thunnissen and Fruytier (2014) illustrated that the scientific staff, including professors, still remain the main actors for early career scientist career development. The ability to attract and retain "top talent" becomes a key issue for universities (Thunnissen, 2015). This responsibility falls generally on the professors. Instead of adhering to any systematic policy (which may or may not exist), professors select their own talent, which increases the risk of favoritism (Van Balen, 2010).

Human Resource Management (HRM) development and talent policies at universities in the Netherlands have taken shape mostly through the development of personalized instruments. The most important example is the so-called postdoctoral tenure-track system, involving a clear career path with predetermined achievement criteria and the Innovational Research Incentives Scheme. However, the actual impact of these instruments on the postdoc population as a whole is quite limited, as their focus is only a very small group of outstanding researchers (Van Arensbergen, 2014).

The second issue is the high amount of ambiguity on the career prospects of postdocs inside and outside of academia (Davis, 2009; Van Arensbergen et al., 2013; Van Balen, van Arensbergen, Van der Weijden, & van den Besselaar, 2012; Van der Weijden et al., 2015). To support young scientists in both academic and alternative career orientations, the Postdoc Career Development Initiative was established in the Netherlands in 2008. The initiative aims to serve as a mediator between PhDs and academia, industry, public–private partnerships, and others (PCDI, 2012). The scope of this initiative has been limited to postdocs in the life sciences.

Despite the importance of the postdocs and their activities, the career development of post-PhDs seems to occur rather coincidentally, as a sort of black box and without the influence of any systematic career planning (Davis, 2009; McAlpine & Emmioğlu, 2014; Van Arensbergen et al., 2013). Generally, policies for postdocs concerning talent management and professional development are immature (Thunnissen, Boselie, & Fruytier, 2013; Thunnissen & Fruytier, 2014) or nonexistent (Van der Weijden et al., 2015). Both Van Balen et al. (2012) and Van Arensbergen (2014) revealed the lack of systematic career policy at universities, which has a crucial impact on talented researchers in making their decision to leave academia.

Van Balen et al. (2012) showed that incidental factors such as "being at the right place at the right time" determine the career path of academics much more than organizational factors such as career policies. Furthermore, multiple investigations showed a lack of systematic career support and planning provided by universities, and revealed the fact that if any support and planning was provided, it had little effect on the career trajectory of postdocs (Van Balen et al., 2012; van der Weijden et al., 2015). The fact that postdoc careers appear to be unstructured and develop on an arbitrary basis makes it hard for them to recognize future opportunities and plan a career (e.g., Chlosta, Pull, Fiedler, & Welpe, 2010).

Several studies indicated that professors are the most powerful actors in the Dutch academia (Meijer, 2002), as they function as intermediaries (Fruytier & Timmerhuis, 1995) and gatekeepers (Meijer, 2002; Van Balen, 2010) between the university and the individual professionals under their support. These professors have an important say in further career building and determine whether or not a prospective academic is admitted to the scientific community. More formalized and externally monitored selection policies cause resistance among established academics. They argue that formal policies would impede professors in their professional freedom and could lead to a false sense of objectivity. However, the lack of formal selection criteria gives the selection committees the freedom to assess the perceived quality of candidates. Because there are no clear criteria to determine the candidate's quality, this approach may lead to the best candidates not getting selected, which results in a waste of high-level academic talent (Van Balen, 2010). Most scientists choose their own ways to avoid any administrative logics and sustain their own autonomy and academic freedom (Teelken, 2015).

A third and related issue involves the organization of guidance and support for postdocs. HRM and talent management for postdocs is relatively undeveloped (Thunnissen, 2015), and postdoc needs in terms of support are largely unknown (O'Grady & Beam, 2011).

The advantages of an involved supervisor are increased awareness of the need to prepare for the future and therefore more "agency" to do so (Scaffidi & Berman, 2011; Davis, 2009). This advantage is accompanied by the fact that postdocs who feel supported are more confident. They also seem to experience less stress about their future (De Boer, 2013; Drost, 2014). This not only leads to a more positive postdoc experience (Scaffidi & Berman, 2011) but will also allow them to be more prepared for a future career (Chen, McAlpine, & Amundsen, 2015; McAlpine & Amundsen, 2016; Scaffidi & Berman, 2011; van der Weijden et al., 2015).

Recent research confirms the importance of guidance and support on an individual level, especially in preparing for the future. Most postdocs do receive guidance from their direct supervisors, but these supervisors usually limit their role to guide the postdoc's professional development. However, guidance on career opportunities, career advice in general, and guidance on private matters such as the work–life balance are very much desired by postdocs (Chen et al., 2015). The possibility for postdocs to prepare for the future by talking about career possibilities with a mentor improves their confidence and reduces the career uncertainty that postdocs experience (Scaffidi & Berman, 2011). Furthermore, a study by Van Balen et al. (2012) showed a direct link between guidance and the career success of postdocs, where having a mentor is a direct indicator of career success when combined with networking.

Therefore, guidance would be a successful way to improve the postdoc experience and the future career chances for postdocs: it improves confidence, increases the chances of acquiring institutional resources, and provides better career chances. Besides those advantages, guidance also improves scholarly performance and leads to a more positive attitude toward their work environment (Van der Weijden, Teelken, de Boer, & Drost, 2015).

SOUTH AFRICA

The first autonomous degree-granting university in South Africa was established in 1873. The number grew to 36 universities over the years, and currently (due to institutional mergers and the induction of "new" universities since 2001) 26 public universities exist (Soudien, 2015). South Africa's higher education arena has been one of constant transformation. Since the country's turn to democratic governance in 1994, at least three key discourses have underpinned change. Du Preez, Simmonds, and Verhoef (2016) argue that transformation in higher education has been (and still is) structural, ideological, and positional.

The structural discourse has been echoed through the *Education White Paper 3* (South Africa, 1997) and the *National Development Plan for Higher Education* (South Africa, 2001). These policies foregrounded initiatives toward redressing the inequalities of South Africa's apartheid past. The focus was on the structural changes needed to transform universities so that they would be more equitable. This focus included equity of access (making higher education more accessible to staff and students), equity of governance (not only within institutions but also at government level), equity of resources (fair distribution and acknowledgment by government), and epistemic equity (the need to include and develop indigenous knowledges).

Parallel to equity was the structural transition toward efficiency. Based on its legacy of fragmented, unequal, and inefficient operationality, the urgency was for higher education to be "planned, governed and funded as a single national co-ordinated system" (South Africa, 1997, p. 2). Most prevalent was to foster these priorities (equity and efficiency transformation) in light of national political, social, and economic transitions with the intention to transform South Africa, not just higher education. At the center of the structural discourse for scholars such as Soudien (2010) was the need to transform a country with a history of racial inequalities and the persistent injustices that this produces.

Inscribed in the *National Development Plan Vision 2030* (South Africa, 2011, p. 264) is the impetus for higher education to be "an inclusive system that provides opportunities for social mobility, while strengthening equity, social justice and democracy." The formulation and release of this 2030 vision by the government's national planning commission was sparked by events such as the racist incident at the University of the Free State (known as the Reitz-sage) and the realization that a structural discourse "limit[s] attention to transformation in higher education to narrow institutional compliance" (Du Preez et al., 2016, p. 2).

An ideology discourse emerged as a broader, more encompassing lens to transformation that engages with various dimensions of injustice and its interplay within higher education. This discourse came under the spotlight through a report by the *Ministerial Committee on Transformation and Social Cohesion and the Elimination of Discrimination in Public Higher Education Institutions* (South Africa, 2008). For Soudien (2010, p. 876) an ideological discourse makes explicit often ingrained beliefs and assumptions, which continue to perpetuate discrimination and injustice. For the government, this approach has necessitated explicit measures (often structural in nature) to hold universities accountable for increased employment and involvement of staff and students that are African, women, and/or from low-income backgrounds (South Africa, 2011). In recent years, this discourse has initiated and necessitated public debates on "epistemological change; discrimination and exclusions in terms of religion, ethnicity, sexual orientation, class and language; Africanisation or decolonisation of the curriculum; beliefs, attitudes, values and commitments of the whole system; power; diversity; and intellectual justice" (Du Preez et al., 2016, p. 3).

In and among the structural and ideological transitions lies the vision of higher education to contribute to national and international positionality. This focus does not undermine or negate the deep history of South Africa but dreams of its future as a leader in teaching–learning and research for the private and public good within and beyond its own borders. One of the national priorities underpinning this dream is the change needed in the qualifications profile of staff from the current 34 percent who hold doctorates to 75 percent by 2030. This change enables "PhD graduates, either as staff or post-doctoral fellows, [to] be the dominant drivers of new knowledge production within the higher education and science innovation system" (South Africa, 2011, p. 267). The internationalization agenda is another such directive which, for the most part, is focused on contributing to the knowledge economy, promoting increased mobility, and recognizing the importance of participating in an integrated world economy (Altbach & Knight, 2007). It entails the process of incorporating international and local dimensions into the core activities of higher education (teaching and research) so as to foster multiple forms of knowledge toward participating as global citizens who find solutions to addressing shared societal issues (Gacel-Àvila, 2005; Qiang, 2003). In a transforming landscape, internationalization is inevitable and higher education cannot escape its participation.

Although they are fixed-term workers, postdocs contribute to South Africa's transforming higher education landscape. Approximately 10 percent will enter academia contributing to the demographic representation as well as inculcating diverse epistemologies. In their capacity as postdocs, their intellectual and scientific projects could be means through which ideological transformation is engaged. The positionality of higher education is transforming with a postdoc population that is local, African, and international. In terms of research innovation, development, and dissemination, postdocs (together with doctoral candidates and academic staff) are regarded as the leaders of new knowledge production, the internationalization of higher education, and skills development for the expectations of 21st century (and post-21st century) research.

South African public universities introduced postdoctoral fellowship programs in the early 2000s. This effort was, in part, underpinned by national priorities geared toward developing the nation's knowledge capital as locally significant and globally competitive. Investing in postdocs to contribute to these priorities was attractive for institutions for three key reasons, the first being cost-effectiveness. Postdocs are classified as fixed-term and need to comply with section 10(1) (q) of the *Income Tax Act of South Africa*, which classify their remuneration to that of a nontaxable bursary. Institutions do not have to provide benefits such as unemployment, pension, and health care. Institutions vary; some institutions will fund this "bursary," while others might enter into partnerships with external funders to cofund or rely solely on external funders. External funders can be local or global as well as private or public. The involvement of these external funders also fuels institutional research budgets and their profile or standing.

A second attraction is the idea of research generation. With the pressure from government for universities to drive a research-intensive agenda, full-time academics are often unable to fully deliver because of their duties in teaching–learning, research, and community engagement. Accordingly, institutions regard research as the primary activity of postdocs. Each institution (and discipline) has their own research niche and focus, and postdocs advance this niche through research activities such as generating scientific research publications, research collaboration through projects and teams, dissemination of research at conferences and other networking opportunities, and building research skills. Postdocs can also be engaged in research-related activities such as postgraduate supervision and education. Some of these research activities generate subsidy, which is another advantage for the institution.

Third, the role of postdocs can cultivate future academics. One-third of South African academics do not hold a doctorate degree. Institutions invest in postdocs as one avenue to produce highly skilled and qualified academics. This investment can take the form of skills development courses wherein postdocs gain expertise in research publications, grant applications, ethics training, postgraduate supervision, and other relevant aspects of academic life. At many institutions, postdocs also engage in teaching in accordance with the *Income Tax Act of South Africa*, which allows for 12 h maximum per week, paid for by the institution. By the time postdocs enter academe full-time, it is envisaged that they will be highly skilled and qualified so as to aid in reaching national priorities such as increased doctoral graduates through their involvement as supervisors with experience (even if this experience is minimal).

To drive these agendas, the postdoc has a profile. Although this profile can be adapted according to context or demand, common features exist. Postdocs are usually young scientists who are within 5 years of obtaining their doctoral degree, often under the age of 45, and, ideally, have obtained their doctoral degree from an institution other than the one granting the postdoctoral fellowship. Postdocs must agree to the duration term as specified by the host. This term has been known to range between 6 months and 3 years. Even if prospective postdocs match this profile, some institutions are also inclined to make a selection based on evidence of or the promise of an individual to carry out the research-related activities being promoted by that institution and the compatibility with the academic host (one person) and/or entity (group) with which the postdoc will be affiliated.

Two common challenges exist for postdocs within this environment: career instability and status (Simmonds and Bitzer, 2017).[2] In general, postdocs do not feel that their fellowship is a form of security into full-time employment (whether in academia or elsewhere). However, in our research with postdocs, there has yet to emerge a postdoc who regrets being part of a fellowship program. Postdocs generally reference some significance, including developing expertise, gaining experience, meeting scholars in their discipline, increasing self-confidence, and so on. For those envisaging an academic career, the postdoc can be especially noteworthy but cannot ensure employability. Postdocs face career instability, not always because of their expertise, but often because of the scarcity of full-time academic positions and the increase (on average 15% annually) in postdoctoral fellows competing for these.

External forces such as economic recessions that affect research funding and support also play a role in career (in)stability. This circumstance has given rise to the troublesome business of prolonged postdocs, where some postdocs hop from one fellowship to the next (either at the same or different institutions) because they are unable to gain permanent employment. Although this sort of employment may provide a living wage, it can have dire consequences for the individual's career trajectory and the country's economy. Postdocs are non-taxpayers; as holders of the highest-level qualification, this employment pattern has implications for the country's skilled workforce. Some postdocs have

---

2. See Simmonds and Bitzer (2017) for a detailed account of the research process, the biographies of the postdocs that participated, and the discussion presented, which was focused on career trajectories and possibilities of postdocs.

advocated for reimaging the postdoctoral fellowship programs offered in South Africa so that postdocs are not trapped in a professional paralysis. Rather, the postdoc should provide its fellows with the competitive edge needed to escape the cycle of prolonged postdoc fellowships.

With regard to status, due to the regulations around remuneration, postdocs are neither permanent staff nor students. This classification carries with it contentions about the status and recognition of postdocs. Postdocs—holders of the respected title "Dr."—often perceive this classification as degrading, an insult to their accomplishments and an attack on their identity. Factors such as these have proven to influence their sense of belonging and productivity at the institution. Being uninformed or less familiar about who postdocs are and what their involvement is has proven to be offensive to some postdocs as it questions their stature. This uncertainty could be because postdoctoral fellowship programs are still a relatively new classification far removed from the more well-known classifications of staff or student. Some institutions still have not put administrative structures in place (such as specifying "postdoc" on identity or access cards), which could aid in making the identification of postdocs known, resulting in frustration. Consequences include an automatic assimilation of postdocs with the general student body, mistreatment in terms of daily operations such as gaining access to research labs, and belittlement when postdocs need to convince administrative staff of their position at the institution.

Although postdocs are assigned to a host (an individual or group), the host or other related parties have been known to interpret the role of the postdoc as additional support for their own research agendas, teaching responsibilities, and related tasks. As such, the postdoc feels obliged to carry out tasks that have not been agreed on in their service agreement. One reason for this situation could be the perception held by some academics that postdocs have an abundance of free time and should serve the larger interests of the department where they are placed. Postdocs have also shared their frustrations that they are often taken less seriously by their peers when they make suggestions for improvements to research or other practices because of their fixed-term status at the institution. These suggestions are seemingly met with the attitude of "what do they know… they have not been here long and will soon be leaving."

Despite these challenges, postdoctoral fellowship programs have proven to contribute positively to South Africa's transforming higher education landscape. This contribution has been a key determinate for South Africa's goal of being recognized as striving toward being a local and global leader of knowledge generation and innovation.

REFLECTIONS ON THE INTERNATIONAL EXPERIENCES OF POSTDOCTORAL SCHOLARS

Understanding the new realities facing postdoctoral scholars requires looking beyond a single disciplinary, institutional, or national context. As illustrated through the four country-specific overviews in this chapter, national systems of higher education vary in how postdoctoral scholars operate within institutional structures. Some countries, such as Kazakhstan and South Africa, are only beginning to see interest in and growth of postdoctoral scholars, while others, such Australia and The Netherlands, have an established practice of postdoctoral experiences as part of higher education systems. Despite these differences in origin, key shared lessons emerge from a comparative overview. First, faculty serve as an important component of the postdoctoral experience. Faculty can be mentors, supervisors, networkers, and colleagues. The lack of standardization related to the faculty–postdoctoral scholar relationship, however, results in unequal treatment and career opportunities. In addition, a consequence of diverse faculty employment conditions is that faculty are not always positioned, empowered, or motivated to adequately support the postdoctoral role. Second, postdoctoral scholars remain poorly situated between doctoral education and more permanent faculty positions, even in higher education systems with a long-standing postdoctoral tradition. Without a more clearly defined role, postdoctoral scholars may get "lost" in academic institutions and not receive the professional development opportunities needed for a robust career pathway. Finally, national systems (and the global landscape of higher education more broadly) seem disinclined to more clearly articulate the postdoctoral role and its significance to academic institutions. As a result, the role can be poorly defined, mismanaged, and misused.

What lessons might observers gain from an analysis of the international postdoctoral experience? One lesson might be the acknowledgment of an uncertain future for the academic profession, characterized by the weakening of tenure and other stabilizing employment constructs as well as decreased autonomy accompanied by increased accountability. The expansion of part-time faculty results in stark variation among the definition of the postsecondary professoriate. Postdoctoral scholars are directly influenced by this uncertain future. As more permanent faculty job opportunities decrease, the competition for these jobs increases. Combined with a higher number of doctoral graduates around the world, an unbalanced supply–demand relationship exists. How this imbalance will be resolved is undetermined.

A related consequence is the increase of prestige, elitism, and credentials. Postdoctoral scholars must not only hold a doctorate but also one in a certain discipline or from a particular type of academic institution. Altbach's (2016) reflections on the center/periphery dichotomy in higher education reverberate on numerous levels. Not only do national systems recruit postdoctoral scholars with training in more mature academic institutions but also progressively, more doctoral students pursue their education abroad before returning to their home countries. Another lesson relates to the balance between the teaching and research function in higher education as well as the role of the postdoctoral scholar in fulfilling these functions. National systems might seek to improve the quality of instruction in the country or expand educational access to previously underserved student populations. If postdoctoral roles privilege research and academic careers in general are perceived to be increasingly unstable, talented individuals might turn away from a career in the classroom.

Ultimately, the country-specific overviews in this chapter offer additional insight into how national policies and economic trajectories directly influence higher education systems as well as the individuals within those systems. A push for neoliberal accountability or an economic effort to reframe industry–education interactions emphasize that postdoctoral scholars do not experience a neutral professional trajectory. The professional trajectory is not solely the result of accumulated individual outputs, talents, and knowledge, but rather the perceived value placed on these components as well as the individual's location within the national and international knowledge economy.

REFERENCES

Adviesraad voor Wetenschaps- en Technologiebeleid. (2014). *Boven Het Maaiveld: Focus Op Wetenschappelijke Zwaartepunten.* [Standing Out from the Crowd: Focus on Scientific Priorities]. The Hague: Adviesraad voor Wetenschaps- en Technologiebeleid.

Aitzhanova, A., Katsu, S., Linn, J. F., & Yezhov, V. (2014). *Kazakhstan 2050: Toward a modern society for all.* Oxford University Press.

Åkerlind, G. S. (2005). Postdoctoral researchers: Roles, functions and career prospects. *Higher Education Research & Development, 24*(1), 21–40.

Altbach, P. (2016). *Global perspectives on higher education.* Baltimore, MD: Johns Hopkins University Press.

Altbach, P., & Knight, J. (2007). The internationalization of higher education: Motivations and realities. *Journal of Studies in International Education, 11*(3/4), 290–305.

Archer, L. (2008). The new neoliberal subjects? Young/er academics' construction of professional identity. *Journal of Education Policy, 23*(3), 265–285.

Auriol, L., Misu, M., & Freeman, R. A. (2013). *Careers of doctorate Holders: Analysis of Labour market and mobility indicators.* OECD Science, Technology and Industry Working Papers, No. 2013/04. Paris: OECD Publishing.

Boyne, G. A. (2002). Public and private management: what's the difference? *Journal of Management Studies, 39*(1), 97–122.

Cannizzo, F. (2017). You've got to love what you do: Academic labour in a cultural of authenticity. *The Sociological Review* (in press).

Cannizzo, F., & Osbaldiston, N. (2016). Academic work-life balance: A brief quantitative analysis of the Australian experience. *Journal of Sociology, 52*(4).

Chen, S., McAlpine, L., & Amundsen, C. (2015). Postdoctoral positions as preparation for desired careers: A narrative approach to understanding postdoctoral experience. *Higher Education Research & Development,* 1–14 (ahead-of-print).

Chlosta, K., Pull, K., Fiedler, M., & Welpe, I. (2010). Should I stay or should I go? Waum machwuchswissenschaftler in der Betriebswirtschaftslehre das Universitassystem verlassen. *Zeitschrift für Betriebswirtschaft, 80*(11), 1207–1229.

Christensen, T., & Lægreid, P. (2001). *New public management.* Aldershot: Ashgate Publishers.

Coates, H., & Goedegebuure, L. (2012). Recasting the academic workforce. *Higher Education, 64*(6), 875–889.

Crossley. (March 19, 2013). *Are PhD graduates expecting too much.* The Conversation. Retrieved from https://theconversation.com/are-phd-graduates-expecting-too-much-11854.

Cummings, W. K., & Teichler, U. (2015). *The relevance of academic work in comparative perspective.* Switzerland: Springer International Publishing.

Davies, B., & Bansel, P. (2005). The time of their lives? Academic workers in neoliberal time[s]. *Health Sociology Review, 14*(1).

Davis, G. (2009). Improving the postdoctoral experience: An empirical approach. In R. B. Freeman, & D. L. Goroff (Eds.), *Science and engineering careers in the United States: An analysis of markets and employment.* Chicago, IL: University of Chicago Press.

De Boer, M. (2013). *Postdoc; nog steeds hopend of echt slopend? Een kwantitatief onderzoek naar het toekomstperspectief van postdocs aan de TU Delft. Masterscriptie.* Amsterdam: Vrije Universiteit.

De Goede, M., Belder, R., & de Jonge, J. (2013). *Academic careers in The Netherlands. Facts & figures.* The Netherlands: The Hagues.

Drost, M. (2014). *Postdocs: tussen wal en schip, met een tijdelijke aanstelling. Masterscriptie.* Amsterdam: Vrije Universiteit.

Du Preez, P., Simmonds, S., & Verhoef, A. (2016). Rethinking and researching transformation in higher education: A meta-study of South African trends. *Transformation in Higher Education, 1*(1), a2. http://dx.doi.org/10.4102/the.v1i1.2.

Fruytier, B., & Timmerhuis, V. (1995). *Mensen in Onderzoek; Het Mobiliseren Van human resources in Wetenschapsorganisaties.* Assen: Van Gorcum.

Gacel-Àvila, J. (2005). The internationalization of higher education: A paradigm for global citizenry. *Journal of Studies in International Education, 9*(2), 121–136.

Gottschalk, L., & McEachern, S. (2010). The frustrated career: Casual employment in higher education. Aust. *Universities' Review, 52*(1), 37–50.

Higher Education Funding Act. (1988). Canberra, Australia: Australian Govt. Pub. Service.

Jumakulov, Z., & Ashirbekov, A. (2016). Higher education internationalization: Insights from Kazakhstan. *Hungarian Educational Research Journal, 6*(1), 38–58.

Kehm, B., & Lanzendorf, U. (2006). *Reforming University Governance: Changing conditions for research in four European countries.* Bonn: Lemmens/Verlag.

Kelly, A., & Burrows, R. (2012). Measuring the value of sociology? Some notes on performative metricization in the contemporary academy. In L. Adkins, & C. Lury (Eds.), *Measure and value* (pp. 130–151). Oxford: Blackwell Publishing.

Kenway, J., Bullen, E., & Robb, S. (2004). The knowledge economy, the techno-preneur and the problematic future of the university. *Policy Futures in Education, 2*(2), 330–349.

Law on Status of Nazarbayev University Nazarbayev Intellectual Schools andNazarbayev Fund. (January 19, 2011). № 394IV. Retrieved from http://www.edu.gov.kz/sites/default/files/zakon_2011_goda_no_394.pdf.

Marginson, S. (2008). Academic creativity under new public management: Foundations for an investigation. *Educational Theory, 58*(3), 269–287.

May, R., Strachan, G., & Peetz, D. (2013). Workforce development and renewal in Australian universities and the management of casual academic staff. *Journal of University Teaching & Learning Practice, 10*(3), 1–23.

McAlpine, L., & Amundsen, C. (2016). *Post-PhD career trajectories. Intentions, decision-making and life aspirations.* United Kingdom: Palgrave Macmillan.

McAlpine, L., & Emmioğlu, E. (2014). Navigating careers: Perceptions of sciences doctoral students, post-PhD researchers and pre-tenure academics. *Studies in Higher Education, 40*(10), 1770–1785.

McWilliam, E. (2004). Changing the academic subject. *Studies in Higher Education, 29*(2), 151–163.

Meijer. (2002). *Behoud talent (keep talent)*. ISBN:978-94-6190-100-2. Retrieved from http://phdacademy.org/wp-content/uploads/2011/08/Behoud_Talent_door_May-May-Meijer.pdf.

Natalier, K., Altman, E., Bahnisch, M., Barnes, T., Egan, S., Malatzky, C., et al. (2016). *TASA working document: Responses to contingent labour in academia*. Melbourne: TASA.

NTEU. (2012). *Casual teaching & research staff survey 2012: Summary of key results*. Retrieved from *www.unicasual.org.au/survey2012*.

OECD. (2017). *Higher education in Kazakhstan 2017*. Paris: OECD Publishing *http://dx.doi.org/10.1787/9789264268531-en*.

Osbaldiston, N., Cannizzo, C., & Mauri, C. (2017). I love my work but I hate my job – early career academic perspective on academic times in Australia. *Time & Society*. http://dx.doi.org/10.1177/0961463X16682516.

O'Grady, T., & Beam, P. S. (2011). Postdoctoral scholars: A forgotten library constituency? *Science & Technology Libraries, 30*(1), 76–79.

PCDI. (2012). *Your PhD as a stepping stone to success. BCF guide*. Retrieved from http://www.pcdi.nl/images/stories/pdf/PCDI_artikel_in_BCF_Guide_2012.pdf.

Qiang, Z. (2003). Internationalization of higher education: Towards a conceptual framework. *Policy Futures in Education, 1*(2), 248–270.

Ruiter, K. (2015). *Postdocs and the future*. Masterthesis BCO. VU University Amsterdam.

Scaffidi, A. K., & Berman, J. E. (2011). A positive postdoctoral experience is related to quality supervision and career mentoring, collaborations, networking and a nurturing research environment. *Higher Education, 62*(6), 685–698.

Scharitzer, D., & Korunka, C. (2000). New public management: Evaluating the success of total quality management and change management interventions in public services from the employees' and customers' perspectives. *Total Quality Management, 11*(7), S941–S953.

Simmonds, S., & Bitzer, E. (2017). The career trajectories of postdocs in their journeys of becoming researchers. In P. du Preez, & S. Simmonds (Eds.), *A scholarship of doctoral education: On becoming a researcher*. Stellenbosch: SUNMeDIA (forthcoming).

Smolentseva, A. (2003). Challenges to the Russian academic profession. *Higher Education, 45*(4), 391–424.

Soudien, C. (2010). Grasping the nettle? South African Higher education and its transformative imperatives. *South African Journal of Higher Education, 24*(5), 881–896.

Soudien, C. (2015). Looking backwards: How to be a South African university. *Educational Research for Social Change, 4*(2), 8–21.

South Africa. (2011). *National development Plan: Vision 2030*. Pretoria: Government Printers.

South Africa, & Department of Education. (1997). *Education white paper 3: A programme for the transformation of higher education*. Government Gazette, 390(18515). Pretoria: Government Printers.

South Africa, & Department of Education. (2001). *National Plan for Higher Education (NPHE)*. Pretoria: Department of Education.

South Africa, & Department of Education. (2008). *The report of the Ministerial Committee into Transformation in Higher Education (MCTHE)*. Pretoria: Department of Education.

Stephan, P. (2012). *How economics shapes science*. Cambridge: Harvard University Press.

Tampayeva, G. Y. (2015). Importing education: Europeanisation and the Bologna process in Europe's backyard—the case of Kazakhstan. *European Educational Research Journal, 14*(1), 74–85.

Teelken, C. (2015). Hybridity, coping mechanisms, and academic performance management: Comparing three countries. *Public Administration, 93*(2), 307–323.

Thunnissen, M. (2015). *Talent management in academia: An exploratory study in Dutch universities using a multidimensional approach* Dissertation. 978-90-393-6268-6.

Thunnissen, M., Boselie, P., & Fruytier, B. (2013). Talent management and the relevance of context: Towards a pluralistic approach. *Human Resource Management Review, 23*(4), 326–336. http://dx.doi.org/10.1016/j.hrmr.2013.05.000.

Thunnissen, M., & Fruytier, B. (2014). Het mobiliseren van human capital: Een overzicht van 25 jaar HRM-beleid op Nederlandse universiteiten. *Tijdschrift voor HRM, 2014*(1), 1–24.

Van Arensbergen, P. (2014). *Talent proof: selection processes in research funding and careers* (Doctoral dissertation).

Van Arensbergen, P., Hessels, L., & van der Meulen, B. (2013). *Talent Centraal. Ontwikkeling en selectie van wetenschappers in Nederland. Rathenau Instituut.* SciSa 1330: The Hague.

Van Balen, B. (2010). *Op het juiste moment op de juiste plaats. Waarom wetenschappelijk talent een wetenschappelijke carrière volgt, Rathenau instituut.* The Hague.

Van Balen, B., van Arensbergen, P., Van der Weijden, I., & van den Besselaar, P. (2012). Determinants of success in academic careers. *Higher Education Policy, 25*(3), 313–334.

Van der Weijden, I., Belder, R., Van Arensbergen, P., & Van den Besselaar, P. (2015). How do young tenured professors benefit from a mentor? Effects on management, motivation and performance. *Higher Education, 69*(2), 275–287.

Van der Weijden, I., Teelken, C., de Boer, M., & Drost, M. (2015). Career satisfaction of postdoctoral researchers in relation to their expectations for the future. *Higher Education, 72*(1), 25–40.

Waaijer, C. J. F. (2016). Perceived career prospects and their influence on the sector of employment of recent PhD graduates. *Science and Public Policy, 44*(1), 1–12. http://dx.doi.org/10.1093/scipol/scw007.

Ward, S. C. (2012). *Neoliberalism and the global restructuring of knowledge and education.* New York and London: Routledge.

World Bank. (2007). *Higher education in Kazakhstan.* Washington, DC: World Bank. http://documents.worldbank.org/curated/en/353061468039539047/Higher-education-in-Kazakhstan.

Index

"Note: Page numbers followed by "f" indicate figures, "t" indicate tables."

A

AAAS. *See* American Association for the Advancement of Science (AAAS)
AAMC. *See* Association of American Medical Colleges (AAMC)
AAMC-GREAT. *See* Association of American Medical Colleges, Group on Graduate Research, Education, and Training (AAMC-GREAT)
Academic/academia, 16, 20, 38, 43–44
    departments, 73–74
    institutions, 204, 211
Academy of Sciences, 209
Accurate information, 72–73
Acting, 127–135
Ad hoc basis, 211
Adaptive perceptions, 153
Adaptive writing perceptions, 153
Agency, 122–124, 127, 149, 185–187
Alumni databases, 63–64
American Association for the Advancement of Science (AAAS), 7–8
Arnold and Mabel Beckman Foundation, 53–54
Asian tigers, 208–209
Association of American Medical Colleges (AAMC), 32–33, 37
Association of American Medical Colleges, Group on Graduate Research, Education, and Training (AAMC-GREAT), 7–8
Australia, postdoctoral scholar experience in, 204–208

B

Beckman Institute in Illinois, 53–54
Benefits, 11, 64, 72–73, 219
    employee benefits packages, 56–57
    fringe, 32
    retirement, 54

BEST. *See* Broadening Experiences in Scientific Training (BEST)
Biomedical Workforce Working Group Report (2012), 18
Blocks, 153, 158–159, 166
Bologna process, 208–210
Broadening Experiences in Scientific Training (BEST), 60–61
Broadening Participation of Groups Under-Represented in Biology, 23–24
Burnout, 149

C

CAAS. *See* Career Adapt-Ability Scale (CAAS)
Canadian funding, 192
Candidate of Science degree, 209–210
Career Adapt-Ability Scale (CAAS), 12–13
Career
    advising, 71, 74–75, 81, 86
    coherence, 122–124
        model, 127
        of STEM Postdocs, 127–135
    context, 122–124, 130–132
    development programming, 58–62
    exploration programs, 64
    gumption, 122–123, 132–134
    integrity, 122–124, 134–135
    literacy, 122–124, 129–130
    outcome, 56, 63
    planning tools, 108
    trajectory, 189–190
"Catch-all" classifications, 4
Central Asian Republic of Kazakhstan, 208–209
"Central" academic institutions, 204
Chairs in resourceful departments, 75–76
Coherent career practice model, 123

College and University Professional Association for Human Resources (CUPA-HR), 37
Committee on Science, Engineering and Public Policy (COSEPUP), 6–7, 10
Conceptual frameworks, 95, 122–124
Consistent conceptual framework development, 167
Consultative strategic policymaking, 74–75
Cornell University, 99
COSEPUP. *See* Committee on Science, Engineering and Public Policy (COSEPUP)
Cross-linguistic methodological instruments creation, 167–168
Cross-national methodological instruments creation, 167–168
CUPA-HR. *See* College and University Professional Association for Human Resources (CUPA-HR)
Cynicism, 149

D

Data collection, 8, 155–157
    and analysis, 126–127
    on Postdocs, 1
Dawkins reforms, 204–205
"Dealbreaker", 111
Departments
    chairs, 75–76
    committed to research strategy, 76
    departmental resources, 81–82
    with doctorates heading for Postdoc positions, 77
        department implementing Postdoc policies, 81f
        expected career hierarchy in STEM departments, 77–81
        factors for not to pursuing faculty positions, 80f
        positions to achieving for graduating doctorates, 78f
        preferred and achieved positions in S&E departments, 79f
        preferred placements for doctorate students, 78f
        research design, 77
    as focal points, 73–74
Dissonance, 99–100
Dutch academia, 214
Dutch academic landscape, 213–214
Dutch system of higher education, 213–214

E

EARLI SIG-REaC. *See* European Association for Research on Learning and Instruction Special Interest Group on Researcher Education and Careers (EARLI SIG-REaC)
Early career researchers (ECRs), 145, 178–179
Effective mentoring, 93–94
Engagement in research, 148
Enhancing Postdoctoral Experience for Scientists and Engineers, 6–7, 29–33
    analysis, 32–33
    overview, 30
    recommendations, 30–32
Entering Mentoring training program, 94
European Association for Research on Learning and Instruction Special Interest Group on Researcher Education and Careers (EARLI SIG-REaC), 169
European context as influence on Postdoc career opportunities, 145–146
European Cross-National mixed-method study, 144
    analysis, 157–158
    descriptive statistics, 155t
    European context as influence on Postdoc career opportunities, 145–146
    European cross-national research program, 146–152
    exploring Postdocs writing perceptions, 154
    Fred's journey plot, 164f
    Geri's journey plot, 163f
    Kelsey's journey plot, 161f
    method, 154–158
    results, 158–165
    strengths and limitations of research approach, 166–169
    writing, 165–166
    written research communication, 152–154
European Cross-National research program, 146–152
    Journey Plots, 150–151
    multimodal interview protocol, 149–150
    Network Plots, 151–152, 151f
    research-related experiences of early career researchers survey, 147–149
European Research Council, 145
Evidence-based strategies
    application of, 106–114
    checklist for Postdoc interviews, 110t–111t

faculty, 106–109
institutions, 112–114
Postdocs, 109–112
Exhaustion, 149
Exploring Postdocs writing perceptions, 154

F

Face-to-face interviews, 126
Faculty, 106–109
faculty–Postdoc interactions, 92
mentor, 92, 97–98, 106–107, 112
Fair Labor Standards Act (FLSA), 32, 37–38,
52–53. *See also* Postdoctoral reform
compliance of institutions with, 38–40, 39f
postdoctoral population affected by, 39f–42f
updates to, 36
direct effects on postdocs, 37–38
institutional response to injunction,
40–42, 41f
including postdocs, 37
Financial duress, 188–189
Finish instruments, 167
FINS-RIDSS interview protocol, 169–172
closure, 172
initial general questions, 170
journey plot instrument, 170–171
network plot, 171
set-up and briefing, 169
FLSA. *See* Fair Labor Standards Act (FLSA)
FoR. *See* Future of Research (FoR)
Foreign postdoctoral, 18, 20
Fred's journey plot, 164f
Future of Research (FoR), 10

G

Geri's journey plot, 163f
Global economic crisis, 144, 177
Graduate Students and Postdoctorates in
Science and Engineering (GSS), 18,
38–39
Grants, 192, 208
GSS. *See* Graduate Students and
Postdoctorates in Science and
Engineering (GSS)

H

High Agency–High Goals, 135–136
High Agency–Low Goals, 136–137
Higher Education Funding Act, 204–205
Higher education sector, 208

Hiring positions, 212
"Holding patterns", 70
"Hub with Spokes" model, 11
Human capital, 91

I

Identity
development, 146
identity-trajectory construction, 179
IDPs. *See* Individual Development Plans
(IDPs)
Incidental factors, 216
Income Tax Act of South Africa,
219–220
Individual Development Plans (IDPs),
12–13, 51
implementation, 56–58
Industrial enterprise research labs, 209
Innate ability, 153, 158
Innovational Research Incentives Scheme, 215
Institutional Research and Academic
Career Development Awards
(IRACDA), 21
Institutional/institutions, 112–114
commitment, 86
policies, 192
postdoctoral office, 49
support, 53
Instrumental labor, 207–208
Intellectual relocation, 184
Interest in research, 148
International Coach Federation, 12–13
International Union, United Automobile,
Aerospace and Agricultural Implement
Workers of America (UAW), 9–10
IRACDA. *See* Institutional Research and
Academic Career Development Awards
(IRACDA)

J

Journey Plots, 147, 150–151, 156

K

K-means cluster analysis, 158
K99/R00 award, 8
Kazakhstan, Postdoctoral scholar experience,
208–213
Kazakhstani Ministry of Education, 210
Kelsey's journey plot, 161f
Knowledge creation, 72, 158

L

Laissez-faire attitude, 165
Learning, 127–135
Legislation reforms, 204–205
Life goals, 188
Likert-type scores, 127
Lingua franca, 168
Linguistic relocation, 190
Logistic regressions, 82–83, 84t
Low Agency–Low Goals, 137–138

M

Maladaptive perceptions of writing, 153
Malfunctioning work environment, 186
MAXQDA data analysis software, 158
MBTI. *See* Myers–Briggs Type Indicator
   (MBTI)
Medical Scientist Training Program (MSTP), 26
Meeting Nation's Needs for Biomedical and
   Behavioral Scientists, 4, 24–27
   analysis, 26–27
   overview, 25
   recommendations, 25–26
Mentees, 93, 96
Mentor, 92, 96
Mentoring, 92–93. *See also* Proactive Postdoc
   mentoring
   characteristics and outcomes of mentoring
      relationships, 93–96
   "mentoring-up" approaches, 113
   models, 93–94
   postdoctoral scholars, 91–92
Merton's Matthew effect, 93
MES funding scheme, 212–213
Misalignment, 101
   Mila's perspective, 102
   Professor Grimes's perspective, 101–102
Mixed methods research design, 124–127
   coding and analysis, 126
   instruments and participants, 124–126
   participant self-identified academic
      discipline, race/ethnicity/status, and
      gender, 125t
   researchers and limitations, 126–127
Mobility, 189–190
   programs, 208–209
Motivation, 185–187
MSTP. *See* Medical Scientist Training Program
   (MSTP)
Multimentored Postdoc, 111
Multimodal interview protocol, 149–150

Myers–Briggs Type Indicator (MBTI), 60
myIDP, 21

N

National budget cycle, 212–213
National Cancer Institute, 19
National Development Plan Vision 2030, 218
National Institute of General Medical Sciences
   (NIGMS), 21
National Institutes of Health (NIH), 1, 32,
   34–35, 50
   NIH BEST Awardee institutions, 61–62
   NIH NPA White Paper, 8
National Postdoctoral Association (NPA), 1,
   7–10, 32, 50, 59t, 121–122
National Research Act (1974), 2–3
National Research Council, 23
National Research Mentoring Network
   (NRMN), 95
National Research Salary Award (NRSA),
   52–53
   program, 2–3, 26
   recommendations, 4
National Science Board, 69–70, 85–86
National Science Foundation (NSF), 1, 18, 32,
   34, 50
Nazarbayev University (NU), 210–211
Neoliberal strategies of governance, 206
Neoliberalism, 205
Netherlands, Postdoctoral scholar experience,
   213–217
Network Plots, 147, 150–152, 151f
New Public Management (NPM), 213–214
NIGMS. *See* National Institute of General
   Medical Sciences (NIGMS)
NIH. *See* National Institutes of Health (NIH)
non-STEM postdocs, 62
NPA. *See* National Postdoctoral Association
   (NPA)
NPA Institutional Survey, 54–55
NPM. *See* New Public Management (NPM)
NRMN. *See* National Research Mentoring
   Network (NRMN)
NRSA. *See* National Research Salary Award
   (NRSA)
NSF. *See* National Science Foundation (NSF)
NU. *See* Nazarbayev University (NU)

O

Online survey, 124
Outcomes-based frameworks, 96

P

Pan-European influence, 145
Parallel to equity, 217–218
Parental codes, 158
Participants, 157
Passionate investment, 184
PDA. *See* Postdoc associations (PDA)
PDN. *See* Postdoc Network (PDN)
PDO. *See* Postdoc offices (PDO)
Perfectionism, 153, 158
"Peripheral" academic institutions, 204
Perpetual Postdoc, 103. *See also* Postdoctoral
        (Postdoc)
    Daniel's perspective, 105
    Professor McKnight's perspective, 104
Personal values, 188
Personnel Needs and Training for Biomedical
        and Behavioral Research, 2–3
PhD graduates, 177–178, 203, 206, 214, 218–219
PhD versions, 147–148
PIs. *See* Principal Investigators (PIs)
Policy
    actors, 181
    knowledge, 190–193
    policy-based recommendations, 146
Post-PhD, 215–216
    period, 182–183
    researchers' writing perceptions, 153
    versions, 147–148
Postdoc. *See* Postdoctoral (Postdoc)
Postdoc associations (PDA), 7–8, 74–75
Postdoc Network (PDN), 6
Postdoc offices (PDO), 7, 74–75
Postdoc trajectories, 176–177
    context, 177–178
    evidence, 178–182
        Characteristics of HIgher Education
            REsearchers, 201t
        Demographic CHaracteristics by
            UNiversity, 199t
        Postdoc CHaracteristics, 200t
        Postdocs in longitudinal study, 198t
        studies, 179–182
    implications, 195–197
    research result, 182–193
        agency, motivation, resilience,
            185–187
        categories of work, 183–185
        hoped-for careers, 189–193
        personal interaction with work, 187–189
        Postdocs work, 182–189
    tenure-track positions, 193–195

Postdoctoral (Postdoc), 9, 49–50, 109–112
    addressing employment challenges,
        51–56
    appointments, 121–122
        analysis, 23–24
        and disappointments, 22–24, 26–27
        overview, 22–23
        recommendations, 23
    Associate, 56
    associations, 58–60
    Career Development Initiative, 215
    degree, 210
    Education in United States, 17–21
    Experience Revisited, 33–35
        analysis, 34–35
        overview, 34
        recommendations, 34
    Fellow, 17–18, 56
    fellowship programs, 20, 219–221
    offices, 50, 55
    population in United States, 3–4
    position, 49, 121–122
    Postdoc–faculty transition, 114
    programs, 21
    projected FY 2017 stipend levels, 52t
    Research Fellowships in Biology, 23–24
    researchers' writing profiles, 159
    salary, 52–54
    Scholar, 56, 203
        academic institutions, 204
        Australia, 204–208
        history, 2–6
        institutional—grassroots, 6–13
        Kazakhstan, 208–213
        Netherlands, 213–217
        reflections on international experiences of
            postdoctoral scholars, 221–223
        South Africa, 217–221
    tenure-track system, 215
    training, 165, 56–58
    unions, 9–10
    work, 182–189
Postdoctoral reform, 16–35. *See also* Fair
        Labor Standards Act (FLSA)
    Enhancing Postdoctoral Experience for
        Scientists and Engineers, 29–33
    Meeting Nation's Needs for Biomedical and
        Behavioral Scientists, 24–27
    Postdoctoral Appointments and
        Disappointments, 22–24
    Postdoctoral Experience Revisited,
        33–35

Postdoctoral reform *(Continued)*
  potential reform efforts, 71–73
  The Invisible University, 17–21
  Trends in Early Careers of Life Scientists,
    27–29
Primary patterns, 135–138
  High Agency–High Goals, 135–136
  High Agency–Low Goals, 136–137
  Low Agency–Low Goals, 137–138
Principal Investigators (PIs), 49, 144, 176–177
Proactive Postdoc mentoring. *See also*
    Mentoring
  application of evidence-based strategies,
    106–114
  case studies, 100–105
    misalignment, 101
    perpetual Postdoc, 103
    reflection, 105
    role models, 102
  characteristics and outcomes of mentoring
    relationships, 93–96
  complexities of real-life research mentoring,
    96–100
  human capital, 91
  mentoring, 92–93
Process-related frameworks, 96
Procrastination, 153, 158
Productive struggler, 159, 163, 165–166
Productivity, 153, 158
Professional development programming, 58–62
Purdue University, 99

Q

Qualitative analysis, 158–165
Qualitative elements, 124
Qualitative subsample, 157
Quantitative analysis, 157–159
  Postdoc researchers' writing profiles, 159
Quantitative elements, 124

R

Randomness factor, 180
Real-life research mentoring, complexities of,
  96–100
Recognition-based efforts, 99
Reflection, 105
  on international experiences of postdoctoral
    scholars, 221–223
Regularly scheduled meetings, 58
Regulation, 149
Reitz-sage, 218

Research
  conceptions, 148
  hypothesis, 74–77
  research-intensive universities, 97–98
  research-related experiences of early career
    researchers survey, 147–149, 155
  Scholar, 56
Research Associate Association, 11, 56
Resilience, 185–187
  emotional, 156–157
  variation and implications, 185–187
Resources, 190–193
Responsibilities, 190–193
Robust qualitative analysis procedures
    creation, 168–169
Role models, 102
  Jenna's perspective, 103
  Professor Smith's perspective, 103

S

S&E fields. *See* Science and engineering fields
    (S&E fields)
"Salary basis test", 37
Scholarly communication, 148
Science, technology, engineering, and
    mathematics fields (STEM fields), 62,
    77, 121–122
  as agents of own career development, 127
  analysis, 135–138
  career coherence, 127–135
    career context, 130–132
    career gumption, 132–134
    career integrity, 134–135
    career literacy, 129–130
  Postdocs, 127–138
  results from survey of postdoctoral scholars
    in, 128t
Science and engineering fields (S&E fields),
    69–70
  analyses, 81–84
  chairs in resourceful departments, 75–76
  departments
    committed to research strategy, 76
    with doctorates heading for Postdoc
      positions, 77
    as focal points, 73–74
  descriptive statistics of study variables, 83t
  Postdoc office and/or Postdoc association, 81
  Postdoc scholars in, 69–70
  potential reform efforts, 71–73
  research hypothesis, 74–77

Science identity, 95
"Science The Endless Frontier", 2–3
"Scut work", 184
Self-assessment
  tools, 60
  workshops, 64
Sigma Xi Postdoc Survey (2005), 58
Social and cultural capital research, 95
Social cognitive career theory, 95
Social support, 148
Socialization, 97
South Africa, Postdoctoral scholar experience,
  217–221
Soviet legacy, 209
Stakeholders, 1, 7–8
STEM fields. *See* Science, technology,
  engineering, and mathematics fields
  (STEM fields)
Stereotype threats, 97–98
Strengths and limitations of research approach,
  166–169
  consistent conceptual framework
    development, 167
  creation of cross-national and cross-
    linguistic methodological instruments,
    167–168
  robust qualitative analysis procedures
    creation, 168–169
Stress, 149
Struggler, 165–166
  productive, 163
  struggler-profile holders, 159
Subsidy, 219
Successful mentoring relationships,
  94–95
"Superpostdocs", 11
Supply-driven model, 204–205
Survey
  of Doctorate Recipients, 24
  findings, 127

T
T32 announcement, 61
Temporary contracts, 214–215
Tenure-track positions, 193–195
Term limit, 56–57, 72

"The Invisible University", 16–22, 28, 30
  analysis, 20–21
  overview, 17–19
  recommendations, 19–20
Three-level degree system, 208–209
Time investment, 75–76
Tracking Postdoc career outcomes, 63–64
Transparency in career outcomes, 28
Trends in Early Careers of Life Scientists, 5,
  27–29
  analysis, 28–29
  overview, 27
  recommendations, 28
  report, 5
Typology-related frameworks, 96

U
UAW. *See* International Union, United
  Automobile, Aerospace and
  Agricultural Implement Workers of
  America (UAW)
University of California system (UC system),
  9–10

V
Variable "research strategy", 81–82
Veni, 145
Vici, 145
Vidi, 145
Visual methods, 150

W
"Welcome to My Lab" letter, 106–108,
  112–113
Western PhD holders, 212–213
Western-style PhD degree, 209–210
Work environment, 177, 185–186, 194
Work–life balance, 188
Writing, 165–166
  perceptions development, 154
  Post-PhD Researchers' Writing
    Perceptions, 153
Written research communication, 152–154
  post-PhD researchers' writing perceptions, 153
  writing perceptions development, 154

Printed in the United States
By Bookmasters